In Praise of
Simple Physics

Oliver Heaviside (1988, 2002)

Time Machines (1993, 1999)

The Science of Radio (1996, 2001)

An Imaginary Tale (1998, 2007, 2010)

Duelling Idiots (2000, 2002)

When Least Is Best (2004, 2007)

Dr. Euler's Fabulous Formula (2006, 2011)

Chases and Escapes (2007, 2012)

Digital Dice (2008, 2013)

Mrs. Perkins's Electric Quilt (2009)

Time Travel (1997, 2011)

Number-Crunching (2011)

The Logician and the Engineer (2013)

Will You Be Alive Ten Years from Now? (2014)

Holy Sci-Fi! (2014)

Inside Interesting Integrals (2015)

In Praise of Simple Physics

The Science and Mathematics behind Everyday Questions

PAUL J. NAHIN

PRINCETON UNIVERSITY PRESS

PRINCETON AND OXFORD

Published by Princeton University Press, 41 William Street,
Princeton, New Jersey 08540

In the United Kingdom: Princeton University Press,
6 Oxford Street, Woodstock, Oxfordshire OX20 1TR

press.princeton.edu

Jacket illustration by Anne Karetnikov

Library of Congress Cataloging-in-Publication Data

Nahin, Paul J., author.
In praise of simple physics : the science and mathematics behind
everyday questions / Paul J. Nahin.
pages cm
Includes bibliographical references and index.
ISBN 978-0-691-16693-3 (hardcover : alk. paper) —
ISBN 0-691-16693-5 (hardcover : alk. paper)   1. Physics.
2. Mathematical physics.   I. Title.
QC21.3.N34 2016
530—dc23   2015031463

British Library Cataloging-in-Publication Data is available

This book has been composed in ITC New Baskerville

Printed on acid-free paper. ∞

Typeset by Nova Techset Pvt Ltd, Bangalore, India
Printed in the United States of America

1 3 5 7 9 10 8 6 4 2

For Patricia Ann

Contents

Foreword

Physics is a glorious brew of diverse ingredients. No single approach is sufficient when grappling with Nature. Experiment and observation are essential, of course, but so, too, are concepts, pictures, imagination, mathematics, and physical intuition, topped off with logical consistency. We are explorers in a maze with mysteries at every turn—not for the faint of heart!

Learning physics and *teaching* physics (two sides of the same coin) are, likewise, all over the map. There are laboratories, lectures, problem sessions, computer calculations, and books—any approach that can help us understand. And books themselves take different approaches. Some are "top-down," starting with physical laws and then developing examples and applications. Others are historically based, developing physics as the author imagines it was invented or might have been invented if real history were not so full of diversions and blind alleys. Yet other books are "conceptually" based, avoiding all mathematics like the plague. And still others are packed with mathematical analysis but are thin on concepts, illustrations, and applications. Each approach has merit.

In *In Praise of Simple Physics*, Paul Nahin takes a different, refreshing slice through the subject. He shows us some really interesting examples of applying simple physical principles to a variety of special cases, questions, and puzzles.

There are a myriad of topics here: we learn about wringing more of our energy from renewable sources in the chapters "Energy from Moving Air" and "Energy from Moving Water." There is the futuristic chapter, "Rapid Travel in a Great Circle Transit Tube." We find out how best to catch a baseball, how to measure gravity in our garage, and why the sky is dark at night. We learn about an error committed by the great Isaac Newton himself. We even learn how to figure out which of three light switches in the basement controls a lightbulb in the attic, with only a single trip up the stairs!

I learned a lot from this book. I have been doing physics and teaching physics for a long time, but there is always more to learn. For example, I have long enjoyed using dimensional analysis to help solve mechanics problems, by requiring consistency in the fundamental dimensions of mass, length, and time in equations. Yet, several of the beautiful examples given in this book I had never seen before!

This book does not pussyfoot its way around analysis. The reader is presumed to know some beginning differential and integral calculus. There is no sweeping math under the rug: if the way to solve the problem lies through doing an integral or two, Nahin does not wave his wand and say, "Now after performing the integration, here is the result." He dives right in and shows you all the details. So if you are already well versed in simple calculus, you can skim these parts while admiring his straightforward, clear development; but if you are tentative or rusty, every step is there for you to read, so you can catch up on what you never quite learned or what you may have forgotten.

If you have read any of Nahin's previous books, you will not be surprised that this one is also chock-full of entertaining, informal, and sometimes surprising examples on many topics. Whether you are a practicing scientist, a layperson with some background in math and physics, or a student at any level (as long as you have some calculus or are willing to learn), you will enjoy delving into the delightful chapters in this book.

T. M. Helliwell
Burton Bettingen Professor of Physics, Emeritus
Harvey Mudd College
Claremont, California
February 2015

Preface with Challenge Problems

Physics should be made as simple as possible.
But not simpler.
—*Albert Einstein*

Teaching thermal physics
Is as easy as a song:
You think you make it simpler
When you make it slightly wrong!
— *Mark Zemansky*[1]

A mathematical argument is, after all, only organized
common sense.
— *George Darwin*[2]

I've made a curious observation about how the typical "person on the street" (assuming that this is actually a meaningful concept) reacts to the announcement of any new, amazing scientific discovery. Usually it's astonishment but, occasionally, the reaction is over-the-top. An example is the announcement a few years ago from a research group at CERN (the famous high-energy particle physics laboratory near Geneva) that they had observed faster-than-light neutrinos. I remember what *I* thought when I heard the breathless report on TV— somebody needs to get their measurement instruments recalibrated! (It turned out to have been a bad cable connection.)

One of my high school acquaintances, however, with whom I still occasionally exchange email, was, to my bemusement, simply bouncing up and down with excitement. Trained as a lawyer who, I suspect, has little understanding of the physical and mathematical arguments underlying special relativity, my correspondent was quite put out with me when I replied to an enthusiastic email with an unenthusiastic view of the CERN report. We repeated the awkwardness the very next year, in 2012—my correspondent the excited cheerleader and I the jaded party pooper—when the possible discovery at CERN of the Higgs boson (the so-called God particle) was announced. *That*, I admit, I was more willing to believe had merit. But I was still puzzled at *why* this intelligent person, who had a very successful career spanning decades as a high-level corporate lawyer, was so willing (indeed, was positively eager) to jump onto the bandwagons of excitement that invariably swell up around every spectacular but preliminary announcement in physics.

Actually, I have to admit that my high school acquaintance is not nearly so scientifically lost as are so many other Americans. In a guest editorial in the *American Journal of Physics* (October 1996), Michael Shermer (author of the 1997 book *Why People Believe Weird Things*) quoted a 1990 Gallup poll that indicated more than half of adult Americans believed in astrology, not quite half believed that dinosaurs and humans lived at the same time, and more than a third believed in ghosts. I suspect those fractions haven't changed much since (or if they have, the change hasn't been for the better). His explanation for this: "[people] cannot accept . . . reality."

Thus, we have the widespread fascination for the items listed in the Gallup poll, along with equally nutty nonsense like the Bermuda Triangle, the Loch Ness Monster, Big Foot, and, of course, the myth that the United States is supposedly hiding an alien spacecraft in the mysterious Area 51 at a top-secret air force base in New Mexico. Hollywood filmmakers love this sort of silly stuff. And why not? It makes them a *lot* of money from the gullible, and many of their science fiction films have done nothing to discourage common beliefs in crackpot "science."[3]

After thinking about this for a while, I've concluded such excitement is generated because these announcements appear to be like magic. If neutrinos can go faster than the speed of light, then, golly, maybe all the neat stuff we oohed and aahed over on *Star Trek* could really happen, such as meeting exotic aliens in other galaxies and traveling backward through time. The depressing corollary I drew is that many people must somehow feel that the everyday world is in some way lacking (or at least deficient) in excitement. That realization made me sad, mostly because it is just so very wrong. The everyday world we live in is already wondrous without any need for wallowing in make-believe. Most people simply take for granted what—if they just knew how to *analytically* think about what they see—are amazing, indeed astonishing, yet completely understandable occurrences.

What my correspondent and those in the same situation lack is knowledge of fundamental physics and mathematics. There is a long tradition in America for educated people to have such knowledge, extending to the earliest days of the Republic. The ideas of Newton, which by the middle of the 1700s were routinely taught in European and English universities, had profound influence on the Founding Fathers. Franklin, for example, actually tried to meet Newton when in London as a young man, and Madison (as a Princeton undergraduate) wrote an essay comparing the world of human affairs with that of nature. And Jefferson's inclusion in the Declaration of Independence of a long section on "natural law" can be directly traced to his reading of Newton's *Principia* and to the writings of others (such as Locke and Voltaire) who also had been similarly influenced.[4]

Let me hasten to assure you that the knowledge I am speaking of is *not* that of a PhD-level theoretical physicist, or that of a math genius possessing an extraordinary ability to manipulate the esoteric symbols of advanced mathematics. Now, obviously, if you are studying what goes on inside a wormhole time machine, or what the universe was like 10^{-10} seconds after the Big Bang, well then, an advanced understanding of general relativity, quantum electrodynamics, and tensor theory would of course be a big help. But that's not the sort of thing we are going to do in this book. The topics discussed in this book will be much closer to home than are either the interior of a wormhole or the details of the stupendously gigantic explosion that was the Big Bang. Instead, we'll be examining things we see (or could arrange to see if we wished to do a little experimenting) in our everyday lives.

Please don't misunderstand me—an advanced understanding of mathematical physics (something, I repeat, that we are *not* going to need *here*) can indeed open some wondrous doors. Some *so* wondrous that I think my high school correspondent would metaphorically explode with excitement (and so deluge me with even more emails). Consider, for example, an essay[5] that appeared nearly 25 years ago, opening with these astonishing words: "Imagine a strange, alien civilization that evolves inside an enormous insulator, which is slowly cooling. Suppose that, unbeknownst to the inhabitants, the insulator will have a transition to a metallic phase below a certain temperature. The inhabitants of this unusual world would, over the course of time, deduce the laws of physics and chemistry. As the insulator cooled, however, it would suddenly become metallic. It would appear to the inhabitants that there would be a sudden change in the basic laws of physics—long-range electromagnetic fields would no longer exist, wave propagation would be altered, etc. Depending on the biological properties of the inhabitants, it is quite probable that the new laws of physics and chemistry would not support life, and the transition would thus be instantly fatal to the civilization. Is there any possibility that a sudden change in the laws of physics could occur in our Universe? Such a question might seem ludicrous if it were not for the fact that, in the standard model of weak and electromagnetic interactions, *such a transition has already occurred!* [emphasis in original essay]."

The authors explained that this change in physical law occurred a long time ago, just after the Big Bang, resulting in the laws we know today. But could such a sudden change happen again? According to one theory discussed in the essay, the answer is yes, if the massless photon we know today suddenly became massive. One consequence of that would be that radio waves would be limited to a range of 1 centimeter! And so, while home cable TV could still work, cell phones, car and airplane radios, and air traffic control radars would not. In arriving at these startling conclusions, the authors traveled through several pages of pretty advanced mathematical physics.

But that is not what we are going to do here. The topics discussed in this book will be far more typical of "real life." The physics required will include such elementary concepts as Archimedes' principle, Ohm's law, Newton's laws of motion, the conservation laws of energy and momentum, calculating the center of mass of a collection of massive bodies, and determining the moment of inertia for simple geometric objects like hollow and solid spheres and cylinders. (When we do use these concepts, I'll remind you of the details as we need them.) The mathematical tools required will be algebra, trigonometry, vectors, and—now and then—even some college freshman calculus. That is, I'll expect you to be aware of material that many bright high school students have mastered before heading off to college. Now, occasionally I *will* extend the math just a bit beyond the freshman level (maybe to the sophomore level), but when I do I will try to be extra gentle in the discussion—and so you may learn some new math here, too, in addition to the physics!

In that spirit, I recall once reading what I call the Julia Child/Rachel Ray definition of physics, according to one high school student whose conversation with a fellow student was overheard by a teacher: "First, take a little algebra and add a bunch of geometry. Then add some more algebra, trig, and some stuff that must be college math. Plus a bunch of things from chemistry that you forgot and even some biology.[6] . . . Mix it all together and you have physics." In keeping with Einstein's dictum, I have tried very hard to keep the discussions simple, but not *so* simple that, as Professor Zemansky lamented, they are simply wrong.

Some readers may be just a bit skeptical at this point and not at all convinced that such elementary tools can actually explain interesting, complex matters. To respond to that concern, here's a dramatic rebuttal. The most highly classified scientific work of the Second World War was the atomic bomb,[7] and any public talk about the nature of such a device was a sure way to get into serious trouble. To appreciate just *how much* trouble, consider what happened after a short science fiction story[8] appeared in early 1944. That story contained an amazingly detailed description of the atomic bomb as a uranium bomb, a U-235 device triggered by a neutron detonator. That was shocking to those in Washington, DC, who were involved in the security apparatus surrounding the Manhattan Project (as the U.S. atomic bomb program was intentionally misnamed). The threat of a security leak was enough to bring both the FBI and the U.S. Army's Counter Intelligence Corps down on the heads of both the author and the editor of the magazine.[9]

Once the war ended, things loosened up a bit, but there were still matters *not to be talked about*. For example, almost immediately after the atomic bombings in Japan in August 1945, Henry Smyth (1898–1986), the head of the physics department at Princeton University, published a book-length report titled *A General Account of the Development of Methods Using Atomic Energy for Military Purposes*. He did this at the behest of General Leslie Groves (1896–1970), head of the Manhattan Project, to serve (it was claimed) the public interest. But not *everything* about the bomb was in the Smyth Report. Indeed, in his introduction to the report, Groves warned readers not to ask for additional information beyond what was printed and threatened all who tried with prosecution under the Espionage Act!

One item that was conspicuously absent was a calculation of the so-called critical mass of a uranium fast-neutron chain-reaction fission bomb (or *gadget,* the euphemism used at Los Alamos for security reasons), that is, the minimum mass of U-235 that would spontaneously explode. Knowledge of the critical mass was crucial to the effort; if it turned out to be too large to be made as a deliverable (by airplane) weapon, then there would simply be no point in building the gadget. It was suggested in the report that such a mass might be anywhere between 1 and 100 kilograms, but the actual value was withheld.

The top theoretical physicist in World War II Germany, Werner Heisenberg (1901–1976)—who was awarded the Nobel Prize in Physics in 1932—severely hindered the Nazi effort in building an atomic bomb through a gross miscalculation of the critical mass for U-235. He thought it would be very large, on the order of *tons.* This error was fatal, in fact, and even with a more than three-year head start over the Americans, the Germans never even got a reactor operational, much less built a bomb. The belief today is that Heisenberg simply never understood how an atomic bomb would actually work, but after the war he found it expedient to claim he made his "mistake" intentionally because of moral objections to developing such a destructive weapon. Most historians of science now believe that to be untrue, a story that Heisenberg spread to both distance himself from his willing support of the Nazi war effort and to "explain" his fundamental physics blunder.[10]

Then, in 1947, a note appeared in the *American Journal of Physics* that showed, using just simple physical arguments and high school math, how to calculate the critical mass to be "weighing about 2.5 kg."[11]

The author of the note, Chinese theoretical physicist Hoff Lu (1914–1997) at the National University of Chekiang, was immune from Groves's threat, since he had worked it all out using just the known laws

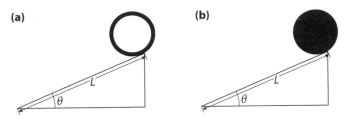

Figure P1. Two cylinders at the start of their race

of physics and math.[12] He had no "inside dope" from anyone involved in the Manhattan Project.

The actual value of the critical mass depends on numerous factors, including the purity of U-235 in the fissionable mass, the density of the mass and its shape, and the nature of the surrounding neutron containment shell (the so-called tamper). Lu's value was certainly closer to the mark than was Heisenberg's.

What Lu did is what we'll do here, although somewhat less dramatically. What I'll do in this book is illustrate how wrong the mathematician G. H. Hardy was when he declared "[Knowledge of] a little ...physics...has no value at all in ordinary life."[13]

Now, to end, here are four quick examples of the level of complexity to the questions we'll examine.

Suppose we have two identically tilted inclined planes, as shown in Figure P1, and two cylinders (made from the same material) with identical radii and masses. One is a hollow, cylindrical, thin-walled shell (a), while the other is solid (b). We can satisfy the requirements by making the solid cylinder shorter than the hollow one. Now, if we release each cylinder at the same instant, so each starts rolling down its respective plane under the influence of gravity, which one gets to the bottom first? What does your intuition tell you? This question is one in which the physics is the same as what we'll encounter when we study the energy in the ocean tides. We'll *analytically* solve this example later in the book, and you can see then whether your intuition was correct. The analytical approach will tell us not only which one wins but also *by how much* the winner beats the loser.

Suppose we have two rigid, straight rods of equal length L. The rods are hinged together at point b, and the bottom rod is hinged to

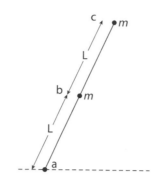

Figure P2. A falling chimney

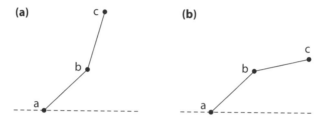

Figure P3. How does the chimney buckle?

the ground at point a, as shown in Figure P2. There are two equal point masses m at points b and c, where the masses of the two rods themselves are insignificant compared with m (and so we'll treat the rods as massless). Starting with the two rods aligned as shown in Figure P2, slightly tilted from the vertical, we then let the structure fall. Does it continue to be aligned as it falls, or does it *buckle* in one of the two possibilities shown in Figure P3? More specifically, *if* buckling occurs, then does it do so as shown in (a) or in (b)? What does your intuition tell you? This question is a simple model of how a tall, slender chimney behaves as it falls over (think of all the TV news videos you've seen of old buildings being demolished with high explosives). We'll answer this question analytically at the same time as we study the first example.

Figure P4 shows two bobsledders, A and B, about to race along two (different) frictionless paths. Each initially has a purely horizontal speed of v_0. A's path is always horizontal, while B's path resembles that

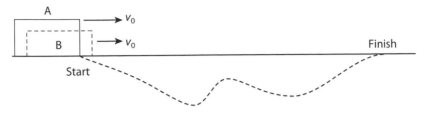

Figure P4. Which bobsledder wins the race?

of a roller coaster but never rises above A's path. Who wins the race? (You'll find the solution at the end of Chapter 1.)

The operator of an automobile traveling on a road with an uphill slope of 8% (the road rises 8 feet for every 100 feet of horizontal displacement) sees a pedestrian in an approaching crosswalk and slams on the brakes. The wheels lock, and the tires leave skid marks that are 106 feet long. The posted speed limit is 25 mph. Was the driver speeding? If, instead, the slope was 8% *downhill*, how would your answer change? (The answers to these questions—see Chapter 4— could have serious legal implications if the pedestrian is hit.)

Now, by giving you these examples of things to come, I have, of course, opened the door to the natural question, just *how* did I select what to include in this book? The everyday world is replete with fascinating physics, after all, and we'd need a much larger book than this one to address even a minute fraction of it all (and a crane to lift it). So, frankly, what's on the following pages *is*, to a large extent, arbitrary, being a compromise between what I personally find interesting and my goal of achieving some sort of representative balance on "everyday physics."

The absence of some topics may be startling to some: there isn't anything, for example, on either the Doppler effect or variable-mass systems, topics that my original table of contents included. They are important topics, to be sure; however, this isn't an encyclopedia of physics but, rather, a sampler of "simple physics." I eliminated Doppler simply because of space considerations, and I excluded rockets losing exhaust mass, raindrops gaining mass falling through a mist, and other variable-mass systems because I came to the conclusion that their "simple physics" would be more complicated than I wanted to address. However, I did include the chapter on a rapid-transit system using an elevated evacuated tube following a great circle route, even though it uses fairly advanced math, because I decided it was just *too* interesting to skip.

I really hated to eliminate variable-mass systems, because I planned to include in that discussion the following funny story concerning the great Scottish physicist James Clerk Maxwell (1831–1879). In a letter to a friend that he sent from the Cavendish Laboratory in Cambridge, England, on February 15, 1878, Maxwell wrote (as a reply to a question from his friend): "I don't know how to apply [the] laws of motion to bodies of variable mass, for there are no experiments on such bodies any more than on bodies of negative mass. All such questions should be labeled 'Cambridge, Mass.' and sent to U.S."

This seemingly odd passage makes perfect sense once you realize that Maxwell was famous for (besides his physics) having a keen sense of humor. What he was actually saying is that queries about applying the laws of motion to bodies of variable mass should *not* be labeled "Cambridge, Mass." and sent to "us." But I see that I *have* included this story, and so all is good.

One of the main goals I had in writing this book was to rebut a commonly held yet completely erroneous belief: "math is just a bunch of theorems, proofs, and boring multiplication tables" (to paraphrase one very wrong-headed assertion I once overheard), and so it can't possibly result in new knowledge but only in tautologies—*tautology* is just a fancy way of saying "going in a circle." For example, if after a long and laborious analysis, all your equations reduce to declaring that $1 = 1$, well, that isn't wrong, but it also isn't new or even interesting! I think you'll find that every one of the chapters in this book is anything *but* a tautology.[14]

The first chapter is specifically designed to be a quick check for you to see whether you have the math you'll need for this book (there's a lot of background physics in there, too), and you should read that chapter next to see how you do. But the following is a simple, quick test of your math.

What's your reaction to the following, once seen on a bumper sticker at a high school sporting event: "We're number $\frac{1}{2}\log_{10} 100$."? If you're puzzled, well then . . . , but if you *laughed*, then you *are* probably all set for the rest of this book.[15]

Notes

1. Mark Zemansky (1900–1981) was an American physics professor at the City College of New York. He was coauthor of the original *University Physics*, a fantastically successful book first published in 1949 and now in its 13th edition, that countless college freshman from the 1950s to the present day think of fondly (or, in some cases, fearfully).

2. The mathematical physicist Sir George Darwin (1845–1912) was the son of Charles Darwin of evolution fame and a professor of astronomy at the University of Cambridge.

3. For an educational book on Hollywood's regrettable infatuation with crummy science, see Tom Rogers, *Insultingly Stupid Movie Physics: Hollywood's Best Mistakes, Goofs, and Flat-Out Destructions of the Basic Laws of the Universe*, Sourcebooks Hysteria 2007. Here's an example of what I'm talking about that isn't in Rogers's book. In the film *Star Wars*, the planet Alderaan is instantly obliterated by the evil henchmen of Darth Vader, using a ray weapon fired from the Death Star. If we assume Alderaan is Earth's twin (same radius and mass), then the energy required to do that is the energy released in the detonation of 5×10^{22} tons of TNT. That's a *lot* of TNT! The only thing that could have made this situation worse would be being told the weapon was powered by a size-D battery. (And I hope reading that won't give future filmmakers any ideas). To learn how to calculate the energy required to disassemble a planet, see my book *Mrs. Perkins's Electric Quilt*, Princeton University Press, 2009, pp. 150–152.

4. See, for example, I. Bernard Cohen, *Science and the Founding Fathers: Science in the Political Thought of Jefferson, Franklin, Adams, and Madison*, W. W. Norton, 1996. A shorter read is the paper by A. B. Arons, "Newton and the American Political Tradition," *American Journal of Physics*, March 1975, ·pp. 209–213.

5. Mary M. Crone and Marc Sher, "The Environmental Impact of Vacuum Decay," *American Journal of Physics*, January 1991, pp. 25–32

6. Simple physics and biology *do* intersect. The classic example is the relationship between metabolism and size for determining how big (and small) living creatures can be. Imagine that there is a characteristic length L for every living creature that "measures its size." Then, the mass of the creature varies as L^3, while its surface area varies as L^2. The internal metabolic heat generated by the creature varies as the mass (as L^3), while the ability to dissipate that heat varies as the surface area (as L^2). Now, $\lim_{L \to \infty} \frac{L^3}{L^2} = \infty$, and $\lim_{L \to 0} \frac{L^3}{L^2} = 0$. This means that creatures that become "too big" will overheat (when you see a

1,000-pound horse standing in a pasture in 30 °F weather it probably isn't at all uncomfortable), while creatures that become "too small" will freeze. (This last point is a fundamental flaw in the 1957 movie *The Incredible Shrinking Man*, a film that Tom Rogers (note 3) overlooks in his excellent book.)

7. There were, of course, numerous super top secret projects during the war, including the Norden bombsight (said to be able to "put a bomb in a pickle barrel from 20,000 feet"), radar and its countermeasures, the artillery-shell proximity fuse, and the breaking of the German Enigma codes. I believe The Bomb was ultimately number one, however.

8. "Deadline," *Astounding Science Fiction*, March 1944, by Cleve Cartmill (1908–1964).

9. You can read what happened next in an essay ("Let's Call It a Hobby") by Murray Leinster, the pen name of William F. Jenkins (1896–1975), in a collection of science fiction stories he edited, *Great Stories of Science Fiction*, Random House, 1951.

10. See Philip Ball, *Serving the Reich: The Struggle for the Soul of Physics under Hitler*, University of Chicago Press, 2014; Jeremy Bernstein, *Hitler's Uranium Club: The Secret Recordings at Farm Hall*, American Institute of Physics, 1996; *Operation Epsilon: The Farm Hall Transcripts*, University of California Press, 1993.

11. The complete fission of 1 kilogram (2.2 pounds) of U-235 releases the energy of 20,000 *tons* of TNT. (See the last example in the epilogue for more on The Bomb.)

12. "On the Physics of the Atomic Bomb," *American Journal of Physics*, November–December 1947, p. 513. Lu's calculation is remarkably similar to what had been done by the American gadget builders several years earlier: see Robert Serber, *The Los Alamos Primer: The First Lectures on How to Build an Atomic Bomb*, University of California Press, 1992, pp. 25–28. The people at Los Alamos had a black sense of humor about their work: Serber mentions that one bomb designed was so huge that if detonated, it would have killed everybody on Earth and so didn't need to be "deliverable." It was code-named *Backyard* because, since it didn't matter *where* it exploded, *that's* where you could set it off!

13. In his 1940 book *A Mathematician's Apology*. Hardy (1877–1947) was one of the greatest mathematicians of the first half of the twentieth century, and his assertion is an example of the ability of even really smart people to say things they might wish later they hadn't.

14. Tautologies aren't limited to mathematics. My favorite example is something a physics grad student (temporarily disoriented from his or her

preliminary PhD oral exam) could easily have blurted out while recovering from the ordeal: "Never before in history have things been more like they are today than they are right now."

15. Here's a more serious, *practical* math/physics question for you to ponder. If you are making a round-trip flight from A to B and then back to A, does a steady wind blowing from A to B increase, decrease, or leave unchanged, the total travel time compared with when no wind is blowing? *Don't guess*—make a mathematical analysis (it's just high school algebra). You can find the answer at the end of Chapter 1.

In Praise of
Simple Physics

1. How's Your Math?

What would life be without arithmetic, but a
scene of horrors?
— *Sydney Smith*[1] *(in a letter dated July 22, 1835)*

In this opening chapter I'll discuss several examples of the kind of mathematics we'll encounter in "simple physics" questions that may (or anyway *could*) occur in "ordinary life." These are questions whose intent, I think, anybody can understand but that require at least some analytical thinking to answer. The math examples are very different from one another, with their only "unifying" (if I may use that word) feature being a progressively increasing sophistication. The central question to ask yourself as you read each example is, do I follow the arguments? If you can say yes, even if you can't initially work through the detailed analysis yourself, then your understanding is sufficient for the book.

Example 1

Our first example of analytical thinking requires *no* formal math but, rather, *logic* and a bit of everyday knowledge (lit lightbulbs get hot). Think about it as you work through the rest of the examples, and, as with the wind-and-airplane problem at the end of the preface, I'll give you the answer at the end of the chapter.

Imagine that you are in a multistory house with three electrical switches in the basement and a 100-watt lightbulb in the attic. All three switches have two positions, labeled ON and OFF, but only one of the switches controls the lightbulb. You don't know which one. All three switches are initially OFF. One way to determine the controlling switch is with the following obvious procedure: Flip any one of the switches to ON, and then go up to the attic to see if the bulb is lit. If it is, you are done. If it isn't, go back to the basement, turn one of the other OFF switches to ON, and then go back up to the attic to see if the bulb is lit. If it is, then the switch you just turned ON controls the bulb. If the bulb is not lit, then the switch that has never been ON controls the bulb. So, you can obviously figure out which switch controls the lightbulb with at most two trips to the attic.

There is, however, another procedure that *guarantees* your being able to make that determination with just *one* trip up to the attic. What is it?

Example 2

This question also requires no real math but, again, logical reasoning (although it does require an *elementary* understanding of kinetic and potential energy). Suppose you fire a gun, sending a bullet straight up into the air. Taking air resistance into account, is the time interval during which the bullet is traveling upward greater than or less than the time interval during which the bullet falls back to Earth? You might think you need to know the details of the air-resistance drag-force law, but that is not so. All you need to know is that air resistance exists.[2] You may assume that the Earth's gravitational field is constant over the entire up-down path of the bullet (it remains the same, independent of the bullet's altitude). As in Example 1, think about this question as you work through the rest of the examples, and I'll give you the answer at the end of the chapter. *Hint*: Potential energy is the energy of position (taking the Earth's surface as the zero reference level, a mass m at height h above the surface has potential energy mgh, where g is the acceleration of gravity, about 32 feet/seconds-squared), and kinetic energy is the energy of motion (a mass m moving at speed v has kinetic energy $\frac{1}{2}mv^2$).

Example 3

This question does require some math, but it's really only arithmetic involving a lot of multiplying and dividing of really big numbers. In the 1956 science fiction story "Expedition" by Fredric Brown (1906–1972), the following situation is the premise. There are 30 seats available on the first rocket ship trip to colonize Mars, with the seats to be filled by selecting at random 30 people from a pool of 500 men and 100 women. What is the probability that the result (as in the story) is one man and 29 women?

We start by imagining the 30 seats lined up, side by side, from left to right, and then we compute the total number of distinguishable (we assume each person is uniquely identifiable) ways to fill the seats *without regard to gender* from the pool of 600 people. That number, N_1, is[3]

$$N_1 = (600)(599)(598)\ldots(571) = \frac{600!}{570!}.$$

Next, if N_2 is the total number of distinguishable ways to fill the 30 seats with exactly one man and 29 women, then the probability we seek is $P = \frac{N_2}{N_1}$. We calculate N_2 as follows:

there are 30 different ways to pick the seat for the one man

and

there are 500 different ways to pick the one man for that seat.

So,

$$N_2 = (30)(500)(100)(99)(98)\ldots(72) = 15,000\frac{100!}{71!}.$$

The *formal* answer to our question is then

$$P = \frac{15,000\frac{100!}{71!}}{\frac{600!}{570!}} = 15,000\frac{(100!)(570!)}{(71!)(600!)}.$$

I use the word *formal* because we still don't have a single number for P.

The factorials in this expression are all huge numbers, numbers that are far too large for direct calculation on a hand calculator (my calculator first fails at 70!). So, to make things more manageable, I'll use Stirling's asymptotic[4] approximation for $n!$: $n! \sim \sqrt{2\pi n}\, e^{-n} n^n$. Then,

$$P = 15,000 \frac{\left(\sqrt{2\pi}\sqrt{100}\, e^{-100}\, 100^{100}\right)\left(\sqrt{2\pi}\sqrt{570}\, e^{-570}\, 570^{570}\right)}{\left(\sqrt{2\pi}\sqrt{71}\, e^{-71}\, 71^{71}\right)\left(\sqrt{2\pi}\sqrt{600}\, e^{-600}\, 600^{600}\right)}$$

$$= \left\{15,000e\sqrt{\frac{(100)(570)}{(71)(600)}}\right\}\left\{\frac{(100^{100})(570^{570})}{(71^{71})(600^{600})}\right\}$$

$$= \left\{15,000e\sqrt{\frac{(100)(570)}{(71)(600)}}\right\}\left(\frac{100}{71}\right)^{71} 100^{29}\left(\frac{570}{600}\right)^{570}\frac{1}{600^{30}}$$

$$= \left\{15,000e\sqrt{\frac{(100)(570)}{(71)(600)}}\right\}\left\{\left(\frac{100}{71}\right)^{71}\right\}$$

$$\times \left\{\left(\frac{570}{600}\right)^{570}\right\}\left\{\left(\frac{100}{600}\right)^{29}\right\}\left\{\frac{1}{600}\right\}.$$

Each of the factors in the curly brackets is easily computed on a hand calculator, and the result is

$$P = 1.55 \times 10^{-23}.$$

The premise in Brown's story is therefore *highly unlikely*. No matter, though, because while it is *so* unlikely as to be verging on the "just can't happen," it is not impossible, and besides, it's a very funny story and well worth a willing suspension of disbelief.[5]

Example 4

Quadratic equations are routinely encountered in mathematical physics (you'll see an example of this in Chapter 9), and here's an example of a quadratic in the form of a type of problem that many readers will recall from a high school algebra class. Readers may take some comfort in learning that it was incorrectly solved by Marilyn vos Savant in her *Parade Magazine* column of June 22, 2014 (but, to her credit, she quickly admitted her slip in the July 13 column after some attentive readers set her straight).

Brad and Angelina, working together, take 6 hours to complete a project. Working alone, Brad would take 4 hours longer to do the project than would Angelina if she did it by herself. How long would it take each to do the project by themselves?

If we denote Angelina's time by x, then Brad's time is $x + 4$. Thus, Angelina's *rate* of clearing the project is $\frac{1}{x}$ per hour, and Brad's rate is $\frac{1}{x+4}$. So, in six hours Angelina finishes the fraction $\frac{6}{x}$ of the project, and Brad finishes the fraction $\frac{6}{x+4}$ of the project. These two fractions must total the finished project (that is, must add to 1), and so $\frac{6}{x} + \frac{6}{x+4} = 1$. Cross-multiplying, we get $6(x+4) + 6x = x(x+4) = x^2 + 4x$ or, $12x + 24 = x^2 + 4x$ or,

$$x^2 - 8x - 24 = 0.$$

The well-known formula for the quadratic equation gives

$$x = \frac{8 \pm \sqrt{64 + 96}}{2} = \frac{8 \pm \sqrt{160}}{2} = \frac{8 \pm 4\sqrt{10}}{2} = 4 \pm 2\sqrt{10}.$$

Since x must be positive, we use the $+$ sign (the minus sign gives $x < 0$), and so $x = 4 + 2\sqrt{10} = 10.32$. Thus, Angelina can do the project by herself in 10.32 hours, and Brad can do the project by himself in 14.32 hours.

The underlying assumption in this analysis is that when working together, Brad and Angelina work independently and without interference. This is not necessarily the case, depending on the nature of the project. For example, suppose "the project" is making a truck delivery. If Brad can drive a truck from A to B by himself in one hour, and if Angelina can drive the same truck from A to B by herself in one hour, how long does it take for the two of them together to drive that same truck from A to B? It's still one hour! An even more outrageous abuse of logic is the belief that if one soldier can dig a foxhole in 30 minutes, then 1,800 soldiers can dig a foxhole in one second!

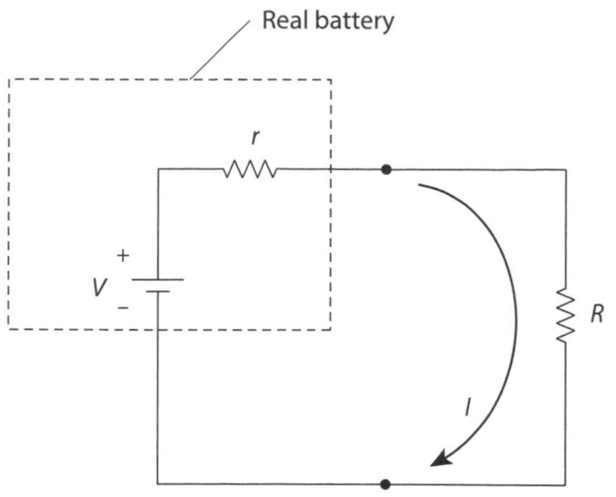

Figure 1.1. What value of R dissipates maximum power?

Example 5

A real battery (with internal resistance $r > 0$ ohms), with a potential difference between its terminals of V volts (when no current is flowing in the battery), is connected to a resistor of R ohms as shown in Figure 1.1. What should R be so that maximum power is delivered to R? This problem is usually solved in textbooks with differential calculus, but that's mathematical overkill, because simple algebra is all that is required.

The current I that flows is (by Ohm's law—see note 1 in Chapter 8 if this isn't clear)

$$I = \frac{V}{r + R}.$$

The power P dissipated (as heat) in R is (where E is the voltage drop across R)

$$P = EI = (IR)I = I^2R,$$

and so

$$P = V^2 \frac{R}{(r+R)^2}.$$

Obviously, $P = 0$ when $R = 0$, and $P = 0$ when $R = \infty$. Thus, there is some R between zero and infinity for which P reaches its greatest value. This value can easily be found with calculus (differentiate P with respect to R and set the result to zero), but all that is needed is algebra. Here's how:

$$P = V^2 \frac{R}{r^2 + 2Rr + R^2} = V^2 \frac{R}{r^2 - 2Rr + R^2 + 4Rr}$$

$$= V^2 \frac{R}{(r-R)^2 + 4Rr} = V^2 \frac{1}{\frac{(r-R)^2}{R} + 4r}.$$

We clearly maximize P by minimizing the denominator of the right-most side of this equation, which just as clearly occurs for $R = r$ (because that makes the first term in the denominator—which is never negative—as small as possible, that is, equal to zero). Thus, $R = r$, and the maximum power in R is $\frac{V^2}{4R}$.

Example 6

In this example you'll see how simple geometry, combined with physics, allows measuring the distance from the Earth to the Moon with fantastic precision. To establish the physics first, all we'll need is the idea that a ray of light incident on a mirror reflects from that mirror at an angle equal to the angle of incidence, as shown in Figure 1.2. This phenomenon was first noted by Euclid, in the third century BC; however, it was not explained until a few hundred years later, in the first century AD, when Heron of Alexandria (in his book on mirrors, *Catoptrica*) observed that the reflection law is a consequence of assuming the ray path *ARB* is the *minimum reflected length path*. That is, if the point R on the mirror was such that $\theta_i \neq \theta_r$, then the resulting total path length would be increased. Heron's observation was the

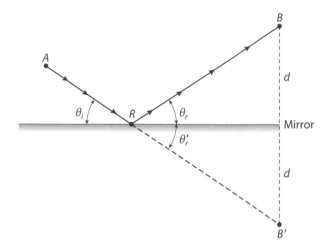

Figure 1.2. Geometry of Heron's reflection law

first occurrence of a *minimum principle* in mathematical physics; such principles play central roles in modern theoretical physics.

Here's a simple geometric proof of Heron's explanation of the mirror reflection law. If B, the destination point, is distance d above the mirror, then B's reflected point (B′) is distance d "below" the mirror. RB and RB' are therefore the equal-length hypotenuses of two congruent right triangles, which means that $\theta'_r = \theta_r$ (referring again to Figure 1.2). Now, the total light path is $AR + RB = AR + RB'$, and this last sum is the path length from A to B′. The shortest path from A to B′ (and so the shortest length for the reflected path, too) is along a straight line, and so $\theta'_r = \theta_i$, which says that $\theta_i = \theta_r$. That's it!

The law of reflection has the following application in an optical device called a *corner reflector* (see Figure 1.3). This gadget allowed the *Apollo 11* astronauts to participate in the 1969 measurement of the distance from the Earth to the Moon to within 2.5 meters! The path of an incoming ray of light to mirror 1 has the vector description (r_x, r_y), and the path of the reflected ray has the vector description $(r_x, -r_y)$.[6] That is, one component of the path vector is reversed, while the other is not; mirror 1, lying along the x-axis, reverses the y-component. The reflected ray continues on to mirror 2, lying along the y-axis, and there the x-component of the path vector is reversed, giving a path vector

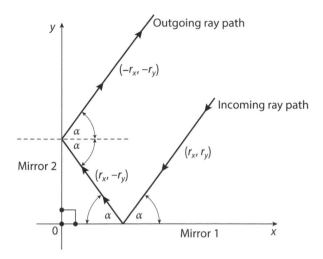

Figure 1.3. A two-dimensional corner reflector

description of the reflected ray off mirror 2 of $(-r_x, -r_y) = -(r_x, r_y)$, which is the *total reversal* of the original incoming ray's path vector. Notice that this means the reflected ray from mirror 2 is *perfectly parallel* to the incident ray on mirror 1, is laterally offset and *reversed* in direction, and these conditions are *independent* of the value of the angle α.

Can the same thing be done in three dimensions? The answer is yes, and that is easily seen once we give the following interpretation to what a reflecting mirror does: the mirror reverses the incident ray's path vector component that is *normal* to the mirror and leaves the other component(s) unaltered. (Look back at the two-dimensional discussion and you'll see that's what happened there.) So, in the case of a *three*-dimensional corner reflector (think of the inside corner of a cube made of three mutually perpendicular mirrors, with the corner of the cube defining the origin of an x, y, z-coordinate system), imagine that mirrors 1, 2, and 3 lie along the xy-, the xz-, and the yz-surfaces, respectively. Then, a ray reflecting off mirror 1 has its z-component reversed, a ray reflecting off mirror 2 has its y-component reversed, and a ray reflecting off mirror 3 has its x-component reversed.

After an incident ray has completed three reflections it emerges from the corner cube reflector in an exactly reversed direction.

The special cases where the incoming ray hits only one (or two) of the mirrors are simply the cases where the incident ray arrives parallel to one (or two) of the mirrors, and so one (or two) of the path vector components happen to be zero (and, of course, the reversal of zero is zero). The *Apollo 11* astronauts placed multiple corner cube reflectors on the Moon's surface, which were then targets for *very* brief (picosecond[7]) laser pulses from Earth. The corner cube reflector sent reflected pulses back to almost precisely where their transmission had occurred, and the elapsed time for the Earth-to-Moon-to-Earth round trip then gave the separation distance. Such measurements have shown that the Moon is *very* slowly moving away from Earth (just an inch and a half per year), and in Chapter 10 you'll learn why.

Example 7

Here's a simple example of high school trigonometry at work in an interesting physics setting. In Robert Serber's book on the U.S. atomic bomb project (see note 12 in the preface), mention is made of the occurrence of the equation

$$x \cos(x) = (1 - a) \sin(x)$$

in one of the theoretical problems studied by the Los Alamos scientists, where a is a given constant. For any particular value of a, what are the *positive* solutions for x ($x \leq 0$ solutions were not *physically* interesting to the bomb designers)?

The most direct way to answer this question is simply to plot both sides of the equation and see where the two plots intersect. This is done for the case $a = \frac{1}{2}$ in Figure 1.4, and we see that the first approximate positive solution is at $x \approx 1.2$, and the next one is at $x \approx 4.6$. There are, of course, an infinite number of positive solutions for the plots beyond $x = 6$ in Figure 1.4. I used a computer to easily generate this figure, but you can appreciate how a technician with just a high school education and a set of math tables could easily make such plots by hand. It would be a laborious task, to be sure, and after a while processing lots of different values for the a-parameter wouldn't be very interesting, but

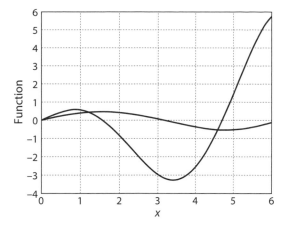

Figure 1.4. Solving $x\cos(x) = \frac{1}{2}\sin(x)$

the Los Alamos scientists had a large number of personnel available who did this sort of thing for them all day long.

Example 8

If pi wasn't around, there would be no round pies!
— *The author, at age 10, has his first "scientific" revelation.*

Everybody "knows" that pi is a number a bit larger than 3 (pretty close to 22/7, as Archimedes showed more than 2,000 years ago) and, more accurately, is 3.14159265.... But *how* do we know the value of pi? It's the ratio of the circumference of a circle to a diameter, yes, but how can that explain how we know pi to hundreds of millions, even *trillions*, of decimal digits?[8] We can't *measure* lengths with that precision. Well then, just how *do* we *calculate* the value of pi? The symbol π (for pi) occurs in countless formulas used by physicists and other scientists and engineers, and so this is an important question.

The short answer is, through the use of an infinite series expansion. For example, we know (after taking freshman calculus) that

$$\int_0^1 \frac{dx}{1+x^2} = \tan^{-1}(x)|_0^1 = \tan^{-1}(1) - \tan^{-1}(0) = \frac{\pi}{4}.$$

TABLE 1.1

Calculating pi *slowly*.

Number of terms	Sum
100	3.1. . . . . . .
1,000	3.14 . . . . . .
10,000	3.141 . . . . . .
100,000	3.1415. . . . . . .

But since

$$\frac{1}{1+x^2} = 1 - x^2 + x^4 - x^6 + \ldots,$$

which you can either derive by doing the implied long division or just confirm by simply cross-multiplying, then

$$\frac{\pi}{4} = \int_0^1 \left(1 - x^2 + x^4 - x^6 + \ldots\right) dx = \left(x - \frac{1}{3}x^3 + \frac{1}{5}x^5 - \frac{1}{7}x^7 + \ldots\right) \Big|_0^1,$$

and so

$$\pi = 4\left(1 - \frac{1}{3} + \frac{1}{5} - \frac{1}{7} + \ldots\right).$$

This famous result[9] is theoretically correct, but, alas, it is also next to useless for *calculating* π because it converges *very* slowly. As the great Swiss-born mathematician Leonhard Euler (1707–1783) wrote (in 1737) about this way of calculating π, to get just 50 digits would be to "labor fere in aeternum" ("work almost forever"). To illustrate that claim, Table 1.1 shows some partial sums for several values of the number of terms used in the sum. As you can see, we have to increase that number by a factor of 10 (!) to determine *each* additional correct digit for pi (the ellipsis dots represent where the sum first fails to give correct digits). We clearly need a series that converges a lot faster (that is, uses far fewer terms to achieve a given number of correct digits).

As it turns out, this is not at all hard to do, as all that is required is a *minor* variation on what we've just done. Writing

$$\int_0^{1/\sqrt{3}} \frac{dx}{1+x^2} = \tan^{-1}(x)|_0^{1/\sqrt{3}} = \tan^{-1}\left(\frac{1}{\sqrt{3}}\right) = \frac{\pi}{6}$$

we have

$$\frac{\pi}{6} = \frac{1}{\sqrt{3}} - \frac{1}{3}\frac{1}{\sqrt{3}}\frac{1}{3} + \frac{1}{5}\frac{1}{\sqrt{3}}\frac{1}{3^2} - \frac{1}{7}\frac{1}{\sqrt{3}}\frac{1}{3^3} + \cdots,$$

and so,

$$\pi = 2\sqrt{3}\left(1 - \frac{1}{3\cdot 3} + \frac{1}{3^2\cdot 5} - \frac{1}{3^3\cdot 7} + \cdots\right).$$

This series converges pretty quickly, and the sum of just the first 10 terms correctly gives the first five digits. The English astronomer Abraham Sharp (1651–1699) used the first 150 terms of the series (in 1699) to calculate the first 72 digits of pi. That's more than enough for physicists!

Example 9

One day a math-deficient frog was sitting on a tiny lily pad
in a big pond—a lily pad that doubled in size each
night—and on this day the pad covered just one-eighth of
the pond. The frog still saw the vast majority of his beloved
water and so was unconcerned.
Then, just three days later, he woke to find the pond had
vanished while he slept.
—*a sad cautionary tale for frogs with their heads in the sand*

Here's a simple application of calculus to a real, present-day concern. Suppose we have a *finite*, nonrenewable resource that is being steadily consumed at an increasing rate. That is, the depletion of the resource is growing exponentially. Specifically, if the quantity of the resource consumed *today* is r_0, and the rate at which it is being consumed increases at a constant rate, then for some k we have

$$r(t) = r_0 e^{kt}, \quad t \geq 0.$$

Such a resource is, for example, oil. If we know r_0, k, and V (how much of the resource is left), then we can calculate how long (T) it will be until the resource is exhausted. The values of r_0 and k are not hard to measure in the case of oil, but the value of V is mostly a guessing game. Just how much oil *is* left in the world? Ten different "experts" will give 10 different answers.

For oil, let's take the present consumption as $r_0 = 6 \times 10^7 \frac{\text{cubic meters}}{\text{day}}$, and $k = 7\%$ per year. Now, no matter what we pick for V there will always be somebody who thinks we are being too conservative. So, let's pick a value that *nobody* could claim to be an underestimate. Let's assume that the *entire planet* is nothing but oil. Nobody could say, then, that there are "undiscovered reserves"! Thus, taking the radius of the earth as 6.37×10^6 meters, we have the volume of the earth as

$$V = \frac{4}{3}\pi \left(6.37 \times 10^6\right)^3 \text{ cubic meters} = 1.083 \times 10^{21} \text{ cubic meters.}$$

That's a lot of oil—but it's still a finite amount—and so we ask: how long until the planet has vanished out the tailpipe of the last car?

The differential amount of oil consumed in differential time dt' is $r(t')dt'$. So, the amount consumed from time $t' = 0$ to $t' = t$ is

$$\int_0^t r(t')dt' = \int_0^t r_0 e^{kt'} dt' = r_0 \left(\frac{e^{kt'}}{k}\right)\Big|_0^t = \frac{r_0}{k}(e^{kt} - 1).$$

At $t = T$ the consumed amount is, by definition, equal to *all* the oil, V, that we started with at $t = 0$, and so

$$V = \frac{r_0}{k}\left(e^{kT} - 1\right),$$

which we easily solve for T by inspection:

$$T = \frac{1}{k}\ln\left(\frac{kV}{r_0} + 1\right).$$

Since $k = 0.07$ per year $= 1.92 \times 10^{-4}$ per day, we have

$$T = \frac{1}{1.92 \times 10^{-4}}\ln\left(\frac{1.92 \times 10^{-4} \times 1.083 \times 10^{21}}{6 \times 10^{7}} + 1\right) \text{ days}$$

$$= (5,208)\ln\left(0.3466 \times 10^{10}\right) \text{ days}$$

$$= (5,208)(21.966) \text{ days} = 114,399 \text{ days} = 313 + \text{ years}.$$

Just three more centuries and the whole planet is gone. Holy cow, this could be bad.

But wait! A returning astronaut has just discovered that there *is* more oil. The Moon! The Moon is all oil, too! Cities worldwide echo with the cheers of car owners who thought they would have to learn how to ride a bike. The world is saved—or is it? What we need to now calculate is, how much more time does the Moon oil extend our ability to consume oil?

Taking the radius of the Moon as 1.74×10^{6} meters, we have the volume of the Moon as

$$\frac{4}{3}\pi\left(1.74 \times 10^{6}\right)^{3} \text{ cubic meters} = 0.022 \times 10^{21} \text{cubic meters}.$$

Thus, starting with the Earth *and* the Moon, we have

$$V = \left(1.083 \times 10^{21} + 0.022 \times 10^{21}\right) \text{ cubic meters} = 1.105 \times 10^{21} \text{ cubic meters}$$

and

$$T = \frac{1}{1.92 \times 10^{-4}}\ln\left(\frac{1.92 \times 10^{-4} \times 1.105 \times 10^{21}}{6 \times 10^{7}} + 1\right) \text{days}$$

$$= (5,208)\ln\left(0.3536 \times 10^{10}\right) \text{ days}$$

$$= (5,208)(21.986) \text{ days} = 114,503 \text{ days}.$$

So, if we consume not only the earth but the entire Moon, too, we'll get an extra 104 days. And then we really will be "outa gas."

The little math story I've just told you reminds me of a funny anecdote told of the great American inventor Thomas Edison. A practical man with little formal education, Edison nevertheless understood the value of education but also never missed a chance to show how a clever person could often work around a technical deficiency. For example, after hiring a young mathematician Edison assigned him the task of determining the volume of a new lightbulb, a bulb designed with an undulating shape. The mathematician carefully reduced the shape to a complicated equation and then laboriously, over a period of hours, integrated the equation over three dimensions to get the volume enclosed. Then, he proudly showed the result to Edison.

Edison congratulated the man on being a fine mathematician, as his computed answer agreed quite well with Edison's own value, which he had arrived at in less than 30 seconds. When the astonished mathematician asked how Edison had done that, the inventor (without saying a word) simply filled the bulb with water and then poured the water out of the bulb into a glass beaker with volume levels marked on the side.

Edison had made his point: math is great, but use it as a tool and not as a crutch.

Solution to Example 1

Flip any switch ON, leave it ON for a minute or so, and then flip it OFF. Then, flip either one of the other two switches ON, and go up to the attic. If the bulb is lit, then the switch you left ON controls the bulb. If the bulb is not lit, feel it. If it's hot, then the switch you turned ON and then OFF controls the bulb. If the bulb is cold, then the third switch (the one you didn't touch) controls the bulb.

This problem, and the Edison lightbulb anecdote, remind me of a goofy "tech joke" that mathematicians like to tell: How many mathematicians does it take to change a lightbulb? The answer is *one*. That's because he or she simply hands the problem off to a group of physicists for whom (claim the mathematicians with lots of snickers and snorts) it is already known that the answer is greater than one. The chief merit of this otherwise outrageous slander of physicists is that it illustrates the powerful trick of reducing an unsolved problem to one whose solution is already known.

Solution to Example 2

On its upward path, the bullet trades kinetic energy for potential energy, as well as irreversibly losing energy because of air drag. So, as the bullet reaches its maximum altitude, it will begin its fall with less potential energy than the kinetic energy it had when it began its upward path. Now, during the fall, at every altitude, its potential energy is equal to what its potential energy was at the same altitude when it was going up. So, the remaining energy (its kinetic energy) at every altitude is less than it was when going up. That is, at every altitude as it falls the bullet is *always going slower* than it was when going upward. So, falling down takes longer than going upward.

Solution to the Bobsledder Problem in the Preface

Looking back at Figure P4, we see that A has a horizontal speed component of v_0 (and no vertical speed component) *at every instant of time*. B, however, has an initial horizontal speed component of v_0 that *increases* whenever he travels downward, because he is accelerated. *Why is he accelerated?* A mass at rest on a horizontal surface exerts a force on that surface, and that surface exerts an equal but opposite (upward) reaction force back on the mass. If the reaction force weren't equal to the downward force, the mass would be accelerated and would *not* be at rest. These comments still hold when the mass moves, but as mass B moves up and down along its curved path, the reaction force has a horizontal component—to the right (which accelerates B) when moving down, and to the left (which decelerates B) when moving up. When B travels upward his horizontal speed component of course decreases back toward v_0, but it is never less than v_0 (remember, no friction). Thus, the horizontal speed component of B is, at every instant, at least as large as A's, and so B wins the race. Notice that this conclusion is true independent of the details of B's path (assuming B's path is what mathematicians call *well behaved*; that is, it doesn't have such sharp angles that B crashes into a wall or jumps free of his path), even though that path is clearly the longer path.

Solution to the Preface Problem in Endnote 14

Let d be the distance between A and B, s the speed of the airplane in still air, and w the speed of the wind. Then, the total round-trip travel time T is the sum of the times spent traveling with, and then against, the wind:

$$T = \frac{d}{s+w} + \frac{d}{s-w} = \frac{d(s-w)+d(s+w)}{(s+w)(s-w)}$$

$$= \frac{2sd}{s^2-w^2} = \frac{2sd}{s^2\left(1-\frac{w^2}{s^2}\right)} = \frac{2d}{s}\left[\frac{1}{1-\left(\frac{w}{s}\right)^2}\right].$$

When there is no wind ($w = 0$) then $T = \frac{2d}{s}$, and when $w > 0$ the denominator in the brackets gets smaller, and we have $T > \frac{2d}{s}$. So, a steady wind *always increases* the total travel time.

Here's a math-free way to see by inspection the special case of $w = s$. In that case the return part of the trip has the plane, with speed s, facing a headwind of the same speed. Thus, the plane *doesn't move* and so will *never* get back to A (that is, $T = \infty$ if $w = s$).

Notes

1. Sydney Smith (1771–1845), an English cleric, was a witty commentator on life in general.

2. All we'll assume is the air resistance drag force law $f(v)$, where v is the speed of the bullet, is *physically reasonable*. That means three conditions hold: (1) $f(v) > 0$ for $v > 0$, (2) $f(v) = 0$ for $v = 0$, and (3) $f(v)$ is monotonic increasing with increasing v.

3. I've written N_1 in factorial notation, where, if n is a positive integer, then $n! = (n)(n-1)(n-2)\ldots(3)(2)(1)$. For example, $4! = 24$. Less obviously, if we notice $n! = n(n-1)!$, then we can conclude that $0! = 1$. Do you see this? (Try $n = 1$.)

4. Named after the Scottish mathematician James Stirling (1692–1770) but actually discovered (in 1733) by the French-born English mathematician Abraham de Moivre (1667–1754). The number e is, of course, one of the most important in mathematics, with the value $2.7182818\ldots$. An asymptotic approximation has the property that while the approximation has an unbounded *absolute* error, its *relative* error approaches zero. That's why we use the $\sim$ symbol and not an equal sign. That is, if $E(n)$ is an asymptotic approximation for some function $f(n)$, then $lim_{n\to\infty}|E(n) - f(n)| = \infty$, but $lim_{n\to\infty}\frac{|E(n)-f(n)|}{f(n)} = 0$. You may say this is more than just arithmetic but, really, you can just look it up in any good book of math formulas and tables.

5. I won't ruin the story for you by revealing where Brown goes with this premise, but if you're wondering, you can find "Expedition" reprinted in *Fantasia Mathematica* (Clifton Fadiman, ed.), Simon and Schuster, 1958. I have long wondered if Brown's story was perhaps inspired by the 1954 hit tune "Thirteen Women and Only One Man in Town," by the great Bill Haley and the Comets (a fantasy about the lone male survivor of a nuclear war).

6. This vector description of the ray path can be thought of as the position vector of an individual photon in the ray.

7. The reason for such brief pulses is the enormous speed of light. Light travels 1 foot in 1 nanosecond, and so 1 inch of travel takes $\frac{1}{12}$ of a nanosecond. To make accurate Moon recession measurements, the timing must then be a *small* fraction of $\frac{1}{12}$ of a nanosecond.

8. Physicists, engineers, and other scientists rarely need to know π to more than five or six digits, so why *trillions*? One example for the *why* comes from those mathematicians who wonder if the digits of pi are uniformly distributed. Crudely, that is, does each of the digits $0, 1, 2, \ldots, 8, 9$ appear 10% of the time "at random"? Mathematicians need those trillions of digits to "experimentally" study this question. (As far as I know, the digits of pi *are* uniformly distributed).

9. It is due to the French mathematician Gottfried Leibniz (1646–1716), who discovered it in 1674. Leibniz was greatly taken by his formula, commenting on it that "The Lord loves odd numbers," obviously ignoring that leading *even* factor of 4.

2. The Traffic-Light Dilemma

The light just turned yellow, so what should I do?
Should I press on the gas or stomp on the brake?
Oh, *Mercy*, I hope I don't make a mistake!
— *the author*

The little jingle above (my sincere apologies to all *real* poets) reflects a quandary that everybody who drives a car faces on a regular basis. Often, the decision is clear, but now and then it is *not* obvious. Or at least it isn't in the short time available to decide, probably no more than a second or so. Should you "go for it" and pray the rear ends of both you and your car get through the intersection before the light turns red, or should you hit the brake pedal and pray your car's front end isn't stopped sticking out into the intersection?[1]

Now, let's first quickly review the simple physics we'll use in this problem. If an object is moving at the constant speed V, then obviously the distance s traveled in the time interval T is $s = VT$. But if that object is *accelerating* at the constant rate of a, then the speed at time $t \geq 0$ is

$$v(t) = V + at,$$

and so the distance covered in the time interval $0 \leq t \leq T$ is

$$s = \int_0^T v(t)dt = \int_0^T (V + at)\,dt = VT + \frac{1}{2}aT^2.$$

Finally, suppose that the object is moving at speed V at time $t = 0$ and then starts to decelerate at the constant rate b. How long does it take to stop the object (to reduce its speed to zero)? The speed of the object is

$$v(t) = V - bt,$$

and so $v(t) = 0$ when $t = \frac{V}{b} = T$. The distance traveled during the deceleration is

$$s = \int_0^T v(t)dt = \int_0^T (V - bt)dt = VT - \frac{1}{2}bT^2 = V\frac{V}{b} - \frac{1}{2}b\left(\frac{V}{b}\right)^2 = \frac{V^2}{2b}.$$

Okay, now we are ready to start.

Clearly, all depends on a number of factors, including how fast you are going, how far it is to the intersection, how much acceleration (and braking deceleration) your car is capable of, how long the light stays yellow, your reaction time, the width of the intersection, and the length of your car. Your brain, which just a moment earlier was mulling over what to have for dinner, has to instantly shift gears and crunch all those factors and quickly decide what to do. Most people intuitively understand that if you are going really fast when the traffic light turns yellow, you can be courting trouble, but the same people are often surprised to learn it is also the case that you can be going *slowly* and still potentially be in difficulty. It's all physics and math (with just a touch of computer graphics) that reveals what is called the all-too-common *stoplight dilemma*.

To start the analysis, let's make the following definitions:

$D =$ width of intersection

$L =$ length of car

$T =$ duration of yellow light

$R =$ driver reaction time

$V =$ speed of car at the instant the light turns yellow

$a =$ car's acceleration under power

$b =$ car's braking deceleration

Next, we consider two cases, A and B. In both cases, the car's front end is distance d from the start of the intersection when the light turns yellow.

Case A: The driver decides to accelerate through the intersection. For this choice to be successful, the *rear* end of the car must be through the intersection before the light turns red. Thus, for the driver to succeed in the attempt,

$$VR + V(T - R) + \frac{1}{2}a(T - R)^2 \geq d + D + L.$$

The meaning of each of the terms in this inequality is as follows. On the left, the first term is the distance traveled before the driver reacts; the second term is the distance traveled after the driver reacts with no acceleration; and the third term is the additional distance traveled, due to acceleration, after the driver reacts. The three terms on the right give the sum of the distance to the intersection, the width of the intersection, and the length of the car, respectively.

Case B: The driver decides to brake to a stop. For this choice to be successful, the *front* end of the car must not enter the intersection. Thus,

$$VR + \frac{V^2}{2b} \leq d.$$

The meaning of each of the terms in this inequality is as follows. On the left, the first term is the distance traveled before the driver reacts, and the second term is the distance traveled as the brakes are applied. The right-hand side is simply the distance to the intersection.

The dilemma occurs when the driver can't satisfy either of the inequalities in A and B. Now, from A and B we have, respectively,

$$d \leq VT + \frac{1}{2}a(T - R)^2 - D - L,$$

and

$$d \geq VR + \frac{V^2}{2b}.$$

A dilemma occurs when *both* of these inequalities are violated; that is, if

$$\frac{V^2}{2b} + VR > d > VT + \frac{1}{2}a(T - R)^2 - D - L,$$

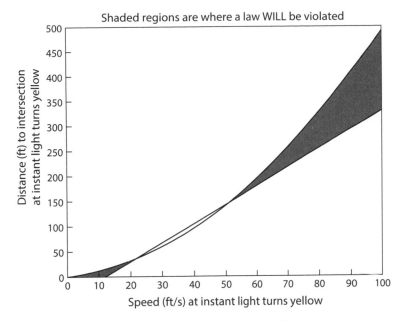

Figure 2.1. The traffic-light dilemma illustrated

then the driver is physically doomed to violating a law. Either he or she *will* run a red light or *will* stop in the intersection when the light is red.

Notice that the left-hand side of the double inequality is parabolic in V, while the right-hand side is linear in V. Thus, if we plot each side with d on the vertical axis versus V on the horizontal axis, then a region that is both below the parabolic *curve* and above the *line* is a region of dilemma, as shown in Figure 2.1 for a typical set of values for the defining variables of the problem:

$D = 45$ feet

$L = 12$ feet

$T = 3$ seconds

$R = 0.75$ seconds

$a = 3$ feet/seconds-squared

$b = 12$ feet/seconds-squared

As you can see, there are *two* shaded dilemma regions. The all-white regions are where one or both sides of the double inequality can be satisfied. The upper white region is where the driver can brake to a stop, and the lower white region is where the driver can accelerate through the intersection. The narrow white region lying between the line and the parabolic curve is where the driver can do either.

Notes

1. This problem periodically reappears in the physics literature and has been around in various forms for a long time. I first encountered it in a paper written over 50 years ago: Howard S. Seifert, "The Stop-Light Dilemma," *American Journal of Physics*, March 1962, pp. 216–218, but the paper I've most closely followed here is by Don Easton, "The Stoplight Dilemma Revisited," *The Physics Teacher*, January 1987, pp. 36–37. A closely related (but not so simple) problem is treated by Seville Chapman, "Should One Stop or Turn in Order to Avoid an Automobile Collison?" *American Journal of Physics*, February 1942, pp. 22–27.

If the "rules" given in this analysis change, then the analysis will have to change, as well. For example, in Arizona an intersection starts at an invisible line defining the extension of a curb, and you are legal as long as the front end of the car has crossed that line when the light turns red. The width of the intersection and the length of the car do not come into play. Similarly, in California, "It is not illegal to deliberately drive through a yellow light. A yellow light means only that the traffic facing the light is 'warned' that a red light will soon follow. As long as your vehicle entered the intersection or passed the crosswalk or limit line before the light turned red, you haven't broken the law." You might try your hand at modifying the analysis I've done here for the Arizona/California rules.

3. Energy from Moving Air

The Betz limit is the best you can do.
—a simple-minded way to best remember Betz

The claim that moving air has lots of energy is hard to dispute once you've been in a severe windstorm or have seen on the news the total destruction caused by a tornado. For a somewhat calmer illustration, at least when viewed from a distance, all you have to do is watch a 250-ton jet aircraft seemingly *float* upward at takeoff. With all the concern these days about energy, it is therefore natural to wonder whether there is any way to tap into the planetwide resource of wind power. The answer is yes.

We can gain an analytical appreciation for the energy of moving air by imagining a mass m of air (think of a cube of air, length s on an edge) that is moving at speed v. The kinetic energy of that mass is $\frac{1}{2}mv^2$. If we further imagine this mass moving through a surface normal to the motion of the mass, then the energy $E = \frac{1}{2}mv^2$ moves through that surface in time interval $\Delta t = \frac{s}{v}$. The *rate* at which that energy passes through the surface—in other words, the *power* P of the wind—is the energy per unit time $\frac{E}{\Delta t}$, and so

$$P = \frac{\frac{1}{2}mv^2}{\frac{s}{v}} = \frac{m}{2s}v^3.$$

That is, the power of the wind varies as the *cube* of the wind speed, and so, for example, a 120 mph wind is *eight* times as powerful as a 60 mph wind (not just *twice*, as one might naively think).

The classic analysis of the optimal extraction of energy from the wind, using a wind turbine, was published long ago, in 1920, by the German engineer Albert Betz (1885–1968). He showed, using simple physics and very elementary mathematics, that a *wind turbine* (in its simplest form, a turbine is simply a cylindrical tube open at both ends, with a bladed fan inside) can convert the kinetic energy of air moving through it into useful power (say, electricity) with a maximum efficiency of 59.3%. This curious value, called the *Betz limit*, can be derived as follows.

To understand how a turbine extracts energy from the wind, imagine air entering the input port of the turbine (with area A) at speed v_i, then encountering fan blades at the reduced speed $v_f < v_i$ (and so exerting a force on the blades, which rotate the shaft of an electric generator), and then finally exiting the output port of the turbine (with area A) at the further reduced speed $v_o < v_f$. The force exerted on the fan blades is the *rate of change* of the moving air's momentum,[1] from when it enters until it leaves the turbine.

If we write ρ as the density of air (in units of kilograms/cubic meter), then the rate at which air mass (kilograms/second) passes through the fan blades is

$$\mu = \rho A v_f,$$

which you can confirm has the units of kilograms/second. This is called the *air flux*. As the air flux enters the input port it carries momentum at the *rate* of μv_i, and as the flux exits the output port it carries momentum at the *rate* of μv_o. Again, you should confirm that μv has the units of momentum per unit time (kilograms·meters/seconds-squared), that is, of force.

Thus, the force on the blades is

$$F = \mu v_i - \mu v_o = \mu(v_i - v_o).$$

Since power is force times speed,[2] the fan power P_f is

$$P_f = F v_f = \mu(v_i - v_o)v_f = \rho A v_f(v_i - v_o)v_f,$$

or

$$P_f = \rho A v_f^2(v_i - v_o).$$

The fan power can also be expressed as the difference between the *rate* at which kinetic energy enters the input port and the *rate* at which kinetic energy exits the output port, and so

$$P_f = \frac{1}{2}\mu(v_i^2 - v_o^2) = \frac{1}{2}\rho A v_f(v_i + v_o)(v_i - v_o).$$

Equating our two expressions for P_f, we have

$$\rho A v_f^2(v_i - v_o) = \frac{1}{2}\rho A v_f(v_i + v_o)(v_i - v_o),$$

or

$$v_f = \frac{1}{2}(v_i + v_o).$$

That is, the air speed at the fan blades is just the average of the speeds at the input and output ports.

Substituting this expression for v_f into either expression (in this case, the first one) for the fan power, we have

$$P_f = \rho A \frac{1}{4}(v_i + v_o)^2(v_i - v_o),$$

or

$$P_f = \rho A \frac{1}{4}(v_i + v_o)(v_i^2 - v_o^2).$$

All the parameters on the right-hand side are either fixed (ρ and A) or out of our control (v_i). We can, however, control v_o, the ouput air speed, with suitable mechanical design of the turbine.

To maximize P_f, we set the derivative of P_f (with respect to v_o) equal to zero:

$$\frac{4}{\rho A}\frac{dP_f}{dv_o} = (v_i^2 - v_o^2) + (v_i + v_o)(-2v_o) = 0$$

or

$$(v_i - v_o)(v_i + v_o) - 2v_o(v_i + v_o) = 0$$

or

$$v_i - v_o - 2v_o = 0$$

or, at last,

$$v_o = \frac{1}{3}v_i.$$

For maximum fan power the air speed at the exit port should be one-third of the input port air speed. Under this condition the maximum fan power, $P_{f\max}$, is given by

$$P_{f\max} = \rho A \frac{1}{4}\left(v_i + \frac{1}{3}v_i\right)\left(v_i^2 - \frac{1}{9}v_i^2\right)$$

$$= \frac{1}{4}\rho A \frac{4}{3}v_i \frac{8}{9}v_i^2 = \frac{32}{108}\rho A v_i^3 = \frac{1}{2}\rho A v_i^3\left(\frac{16}{27}\right).$$

Now, since $\frac{1}{2}\rho A v_i^3$ is the power level at the input port,[3] then

$$\frac{P_{f\max}}{P_{\text{input}}} = \frac{16}{27} = 0.593,$$

a value called the *Betz limit*.

Now, what kind of power levels are we talking about with a reasonably sized turbine? As an example, suppose we have a turbine with a circular input port that is 100 feet in diameter, operating in a 20 mph wind. We have just shown that

$$\frac{P_{\text{input}}}{A} = \frac{1}{2}\rho v_i^3$$

and so, using MKS (meters/kilograms/seconds) units to measure v_i in meters/second and A in square meters, $\frac{P_{\text{input}}}{A}$ will have units of watts/square meter (of input port area). With the density of air at sea level as $\rho = 1.22$ kilograms/cubic meter, we have

$$\frac{P_{\text{input}}}{A} = 0.61 v_i^3 \text{ watts/meters-squared},$$

or using D as the diameter of the circular input port (measured in meters),

$$P_{\text{input}} = 0.61\pi \frac{D^2}{4} v_i^3 \text{ watts} = 0.479 D^2 v_i^3 \text{ watts.}$$

We can convert to the English units more familiar to American readers (that is, to mph for v_i and feet for D) as follows. Since 1 meter is 39.37 inches = 3.28 feet, and since

$$1 \text{ mph} = \frac{5280 \text{ feet}}{3600 \text{ seconds}} \times \frac{12 \text{ inches}}{\text{foot}} \times \frac{1}{39.37 \frac{\text{inches}}{\text{meter}}} = 0.447 \frac{\text{meters}}{\text{second}},$$

which means that

$$1 \frac{\text{meter}}{\text{second}} = \frac{1}{0.447} \text{ mph} = 2.24 \text{ mph.}$$

Then, with D measured in feet and v_i measured in mph,

$$P_{\text{input}} = 0.479 D^2 v_i^3 \frac{(2.24)^3}{(3.28^2)^3} \text{ watts} = 4.3 \times 10^{-3} D^2 v_i^3 \text{ watts.}$$

Thus, with $D = 100$ feet and $v_i = 20$ mph,

$$P_{\text{input}} = 4.3 \times 10^{-3} \times 100^2 \times 20^3 \text{ watts} = 344 \text{ kilowatts.}$$

Thus, for our assumed wind turbine the "Betz" we can hope for (assuming a 100% conversion efficiency of mechanical energy to electrical energy) is

$$P_{f\text{max}} = 0.593 \times 344 \text{ kilowatts} = 204 \text{ kilowatts.}$$

To get an idea of what this means in everyday terms, a modern home with a 200-ampere/110-volt electrical service has a maximum power requirement of 22 kilowatts.

There is one additional, quite interesting calculation we can do, based on our result that the power level of the wind is $\frac{1}{2}\rho A v^3$. For an electric automobile traveling through still air at speed v, the effective "wind" is at speed v, and it produces a drag force on the vehicle. The car's onboard battery must be able to provide sufficient power to overcome this drag force, a power usually written as $\frac{1}{2}\rho A v^3 C_D$, where C_D is a dimensionless *drag coefficient* that takes into account any streamlining the car's shape may offer. For most cars, C_D is about $\frac{1}{2}$. So, to overcome air drag the battery must be able to provide a power output of

$$P = \frac{1}{4}\rho A v^3.$$

If we take A equal to 3 square meters for the projected frontal area of a good-sized car, and $v = 50$ mph (22.3 meters/second), then

$$P = \frac{1}{4}(1.22)(3)22.3^3 \text{ watts} = 10{,}150 \text{ watts}.$$

The voltage for present-day electric car battery packs is typically between 300 and 400 volts, so the required steady current from the battery to overcome air drag at 50 mph is between 25 and 34 amperes. If the car is to have a range of 100 miles, then at 50 mph the battery must be able to supply this current for 2 hours. The total energy required to overcome air drag at 50 mph for 100 miles is therefore[4]

$$10{,}150 \frac{\text{joules}}{\text{second}} \times 3{,}600 \frac{\text{seconds}}{\text{hour}} \times 2 \text{ hours} = 73 \cdot 10^6 \text{ joules}.$$

This quantity is comparable to the chemical energy in one gallon of gasoline. From these values you can understand why *the* crucial issue for the future of electric cars is the development of compact, easily rechargeable batteries that can both store large amounts of energy and then deliver that energy to the car's motor at a rate in the multikilowatt range. (Note: 1 watt = 1 volt × ampere.)

I'll end this chapter with some comments on the emphasis I've put on units, something I'll do throughout this book. Physicists deal with quantities that literally span everything in the universe, from the smallest to the largest, and in doing so they use more than one system of units. To comfortably switch among these various systems is a skill not commonly appreciated outside the sciences, and I was reminded of that one evening while driving home when I happened to come across a radio ad from a dealer in precious metals (gold and silver). The pitchman claimed all investors should have a pile of metal coins stashed in their basement[5] ("Silver could reach $50 dollars an ounce by the end of the year—don't be left out!"). The pitch was to get you to call in for a colorful report (and purchase forms), and to show the dealer's sincerity (whatever that meant), if you did, he would send you a 1-gram bar of silver.

Well, what's that worth? There are 454 grams to a pound and, of course, 16 ounces to a pound, so 1 ounce of silver is 28.4 grams of silver. If silver does reach $50 an ounce, then 28.4 grams will be worth $50, and so that 1-gram silver bar will be worth $1.76—that is, approximately the price of three first-class stamps (in 2016). I'd rather have the stamps, as you can actually *do something* with them.

Notes

1. Momentum is mass times velocity, mv, and the force F is given (for the case where m is a constant) by the formula $F = \frac{d(mv)}{dt} = m\frac{dv}{dt} = ma$, where a is the acceleration (Newton's so-called second law of motion). The units of force, in the metric system, are therefore kilograms · meters/seconds-squared.

2. You can see this dimensionally by writing work (or energy) = force times distance and so $\frac{\text{energy}}{\text{time}} = \text{power} = \text{force} \left(\frac{\text{distance}}{\text{time}}\right) = \text{force times speed}$. This also explains where the familiar "kinetic energy is $\frac{1}{2}mv^2$" comes from. With $F = ma = m\frac{d^2x}{dt^2}$, and $F\frac{dx}{dt} = \text{power} = \frac{dE}{dt}$, where E is energy, we have $\frac{dE}{dt} = m\frac{d^2x}{dt^2}\frac{dx}{dt} = \frac{d}{dt}\left\{\frac{1}{2}m\left(\frac{dx}{dt}\right)^2\right\}$, and so $E = \frac{1}{2}m\left(\frac{dx}{dt}\right)^2 = \frac{1}{2}mv^2$.

3. To be clear on this, air with density ρ enters the input port through area A at speed v_i. The kinetic energy of this moving air, per unit mass, is $\frac{1}{2}v_i^2$, and the mass flow rate is $\rho A v_i$. So, the kinetic energy per unit time, that is, the input power, is $P_{\text{input}} = \frac{1}{2}v_i^2 \rho A v_i = \frac{1}{2}\rho A v_i^3$.

4. The MKS unit of energy is the *joule*, where $1\frac{\text{joule}}{\text{second}} = 1$ watt.

5. When I heard that I was reminded of the old comic book character Scrooge McDuck (the maternal uncle of Donald), who loved to swim in his "three cubic acre" Money Bin. A cubic acre has the units of length to the sixth power (l^6), and that should be a big clue that we have left the real world!

4. Dragsters and Space Station Physics

> If everything seems under control, you're not going
> fast enough.
> —*Mario Andretti, Formula One World Champion driver*

The sport of drag racing is one of raw power. Forget about masterful pit-crews changing four tires in less time than it takes most people to walk twice around their car, and skilled drivers able to perform at ultra-alert levels of awareness for hours under extreme physical stress. A drag race isn't the Indy 500. A regulation drag race, over a measured and timed quarter mile (1,320 feet), is history in just a few seconds. For the most powerful cars, from start to finish, a race takes less than seven seconds and the "only" thing a driver has to do is hang onto the steering wheel and drive a straight line, all the while strapped into a screaming, smoking machine accelerating from a standing start to perhaps 220 mph and even higher.

If a car goes from a standing start to a point s feet away in t seconds, and if we assume a constant acceleration a, then $s = \frac{1}{2}at^2$. So, with $s = 1,320$ feet and $t = 6$ seconds, we have $a = 73.3 \frac{\text{feet}}{\text{seconds squared}}$. Since 1 g (one gee) is $32.2 \frac{\text{feet}}{\text{seconds squared}}$, our driver's assumed uniform acceleration is 2.28 gees, and he or she will feel as if someone weighing more than they do is sitting in their lap. While this result is impressive, it isn't the ultimate. A special class of cars called *rail dragsters* are the fastest-accelerating cars in the world. These machines, weighing over a ton, can blast through a quarter mile in less than 4 seconds and reach speeds exceeding 325 mph with accelerations of over 5 gees.

Drag racing is pure speed, with the winner determined by just two parameters: the elapsed time and the final speed as the finish line is crossed. A multitude of variables go into determining those two parameters, such as the obviously crucial ones of car weight and engine horsepower, as well as tire size, tire pressure, tire friction with the road surface, and many other factors. So, finding a simple formula that predicts how a given car will perform is the dream of the dragster mechanic, and in the late 1950s and early 1960s, the automotive journalist Roger Huntington (1926–1989) succeeded, at least empirically.

Huntington studied the actual performance of many drag racing cars and, after much number crunching, came up with the following rule for predicting the final speed (denoted by MPH):

$$\mathrm{MPH} = 225 \left(\frac{\text{engine power}}{\text{car weight}} \right)^{1/3}$$

where the engine power is measured in horsepower,[1] the car weight is in pounds, and MPH (no surprise here!) is in miles/hour. In 1964 the physicist Geoffrey Fox discovered how to derive Huntington's rule from simple physics, and I'll show you here what he did.[2]

Fox wrote the *mass* (not the weight) of the car as m, the engine power as the constant P, and its speed at time t as v. The kinetic energy of the car at time t is $\frac{1}{2}mv^2$, and this must be the total energy developed by the engine over the time interval 0 to t (if we assume all the engine's energy goes into kinetic energy, and so we ignore the noise and heat energy produced by a roaring monster, as well as the rotational energy in such parts as the wheels, the crankshaft, the clutch, and the pistons). That is,

$$\frac{1}{2}mv^2 = \int_0^t P\,dt' = Pt,$$

because P is a constant. Thus, solving for v, we have

$$v = \sqrt{\frac{2Pt}{m}} = \sqrt{\frac{2P}{m}}\,t^{1/2}.$$

If the duration of the race is T, and since at the end of the race $v = \mathrm{MPH}$, then

$$\mathrm{MPH} = \sqrt{\frac{2P}{m}}\, T^{1/2},$$

or

$$T^{1/2} = \mathrm{MPH}\sqrt{\frac{m}{2P}}.$$

The distance traveled by the car at time $t = T$ is, by definition, $s = 1{,}320$ feet, and so

$$s = \int_0^T v\, dt = \sqrt{\frac{2P}{m}} \int_0^T t^{1/2} dt = \sqrt{\frac{2P}{m}} \left(\frac{2}{3} t^{3/2}\right) \Big|_0^T = \frac{2}{3}\sqrt{\frac{2P}{m}}\, T^{3/2}$$

$$= \frac{2}{3}\sqrt{\frac{2P}{m}} \left\{\mathrm{MPH}\sqrt{\frac{m}{2P}}\right\}^3 = \frac{2}{3}\sqrt{\frac{2P}{m}}\, \mathrm{MPH}^3 \frac{m}{2P}\sqrt{\frac{m}{2P}}$$

$$= \frac{2}{3}\frac{m}{2P}\mathrm{MPH}^3 = \frac{m}{3P}\mathrm{MPH}^3.$$

Thus,

$$\mathrm{MPH} = (3s)^{1/3} \left(\frac{P}{m}\right)^{1/3},$$

which has the same form as Huntington's empirical rule.

Fox's formula, as it stands, using MKS units, gives a speed in meters/second when we insert s in meters, P in watts, and m in kilograms. Dragster mechanics, however, want to insert s in feet, P in horsepower, and m in pounds. So, we get meters/second on the left in Fox's formula, for a quarter-mile race, by writing

$$\mathrm{MPH} = \left(3 \times 1{,}320 \times \frac{1}{3.28}\right)^{1/3} \left(\frac{746P}{w/2.2}\right)^{1/3},$$

because there are 3.28 feet in a meter, 746 watts in a horsepower, and (on the surface of the Earth) the weight of a 1-kilogram *mass* is

2.2 pounds. That is,

$$\text{MPH} = \left(\frac{3 \times 1{,}320 \times 746 \times 2.2}{3.28}\right)^{1/3} \left(\frac{P}{w}\right)^{1/3} = 125.6 \left(\frac{P}{w}\right)^{1/3},$$

where MPH is in meters/second. To convert to miles/hour, we use the conversion

$$1 \text{ meter/second} = 2.237 \text{ mph},$$

and so

$$\text{MPH} = 125.6 \left(\frac{P}{w}\right)^{1/3} \text{ meters/second} = 2.237 \times 125.6 \left(\frac{P}{w}\right)^{1/3}$$

$$= 281 \left(\frac{P}{w}\right)^{1/3} \text{ mph.}[3]$$

Fox's theory seems to be in the ballpark with Huntington's experimental result, but as Fox himself wrote, "Although the discrepancy between [281 for theory and 225 for experiment] doesn't appear to be large, if one cubes it, one finds that about 50% of the theoretical power is wasted." That is, cubing Huntington's formula, which describes how a real drag racer actually performs, predicts a smaller MPH than does Fox's theoretical formula, and the two P's (the rated P and the effective P) are in the ratio

$$\frac{P_{\text{Huntington}}}{P_{\text{Fox}}} = \frac{225^3}{281^3} = 0.51.$$

In his paper Fox explores more complete theoretical descriptions of drag racing, but I'll quit here on the subject while the physics is still "simple."

But before leaving this chapter let me say just a bit more on the distinction I mentioned earlier between weight and mass. (This will also give you some additional examples of converting between MKS and English units.) Mass is a measure of *quantity of matter*—literally, the number of atoms we are dealing with. That number doesn't change as we move the same hunk of matter from one gravitational environment (the surface of Earth) to another (outer space). What does change, however, is the weight of the hunk, which is the gravitational force it experiences, as given by Newton's famous equation $F = ma$ (it is fitting that the MKS unit of force is the *newton*). On the surface of Earth $a = g = 9.8$ meters/seconds-squared, and $F = mg$ is the weight, but in orbit around the Earth where the effect of gravity vanishes (more on this later in the book), we have $F = 0$, and the hunk is said to be weightless. On the surface of the Earth a 1-kilogram mass weighs 2.2 pounds, which is 9.8 newtons, and so the conversion between newtons and pounds is 1 newton = 0.225 pounds.

Sixty years ago a quite interesting note appeared in the *American Journal of Physics* that nicely illustrated the distinction between mass and weight, an illustration that was borderline science fiction at the time but that has since become a routine experience for astronauts onboard the International Space Station.[4] We were asked to "Suppose yourself to be working on a space station and expected to manipulate a ten ton mass of some kind. You are 'standing on' the outer surface of the station (of extremely large total mass), in front of an unyielding wall. The mass is approaching you with a velocity of 1 foot/second, threatening to crush you against the wall. The primary question is: can you expect to stop it, or should you 'evacuate the premises!'"

The author ended his note with the claim, "assuming only that the stopping process occurs under uniform deceleration and takes three feet (linear distance, not a freak of nature) . . . the job is well within the physical capabilities of a normal person, a force of about 100 pounds acting for six seconds being sufficient." He didn't show the calculations behind this claim, but here's how he may have reasoned.

First, the reference to a "10 ton mass," that is, a hunk of matter weighing 20,000 pounds on the surface of the Earth, would be better (I think) described as a

$$\frac{20{,}000 \text{ pounds}}{2.2 \text{ pounds/kilogram}} = 9{,}091 \text{ kilograms}$$

mass.

Now, a mass experiencing a steady deceleration of a, from an initial speed V at time $t = 0$, has the speed

$$v = V - at$$

and so is reduced to zero speed at time $t = T$, where

$$T = \frac{V}{a}.$$

During the deceleration the mass travels distance

$$D = \frac{1}{2}aT^2 = \frac{1}{2}a\frac{V^2}{a^2} = \frac{V^2}{2a},$$

and so

$$a = \frac{V^2}{2D},$$

which gives

$$T = \frac{V}{\frac{V^2}{2D}} = \frac{2D}{V}.$$

Since $D = 3$ feet and $V = 1$ foot/second, we have $T = 6$ seconds.

Further, since $F = ma$ we have

$$F = m\frac{V^2}{2D} = 9{,}091 \text{ kilograms } \frac{\left(1\frac{\text{foot}}{\text{second}} \times 1\frac{\text{meter}}{3.28\,\text{feet}}\right)^2}{2 \times 3\,\text{feet} \times 1\frac{\text{meter}}{3.28\,\text{feet}}}$$

$$= 462 \frac{\text{kilogram} \cdot \text{meter}}{\text{seconds squared}}$$

$$= 462 \text{ newtons} \times 0.225 \frac{\text{pounds}}{\text{newton}} = 104 \text{ pounds},$$

just as the AJP author claimed.

Finally, this is a good place to solve the final challenge question I gave you in the preface, concerning the skidding car. We will use just the simple physics we've already discussed, plus the very simplest ideas on friction. Friction is, *in detail*, a complicated physical process, but for our purposes here we can use a very elementary mathematical description of it that still gives pretty good answers.

It is experimentally found that if a mass m moves at speed v over a level surface, there is a retarding force f (for friction) given by μmg, where μ is a positive constant called the *coefficient of friction*. To a first approximation, μ is independent of both m and v. The quantity mg

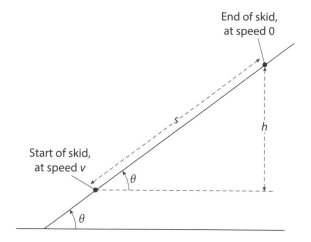

Figure 4.1. The geometry of a skid of length s

is a force (the weight) perpendicular to the surface, and in the more general case of a surface inclined at angle θ to the horizontal, the force perpendicular to the surface is $mg \cos(\theta)$. For the case of a rubber tire moving on a concrete road surface, the value of μ is experimentally found to be greater if the tire is rotating than it is if the tire is *not* rotating (that is, the tire is *skidding*).[5]

In Figure 4.1 we have a mass m skidding on an uphill sloping road at angle θ through a distance s until the mass comes to a stop. (For the second part of the preface problem, in which the skidding is on a *downhill* slope, $\theta < 0$.) The vertical rise of the mass from start to finish of the skid is h. If the speed of the mass at the start of the skid is v, then when the mass comes to a stop at the end of the skid the mass has lost kinetic energy but has gained potential energy. The net energy *loss* is

$$\frac{1}{2}mv^2 - mgh$$

which must be the energy dissipated by the frictional force acting over the length of the skid. So, since energy is "force times distance" (see note 2 in Chapter 3), we have[6]

$$\frac{1}{2}mv^2 - mgh = fs = \mu mg \cos(\theta)s.$$

Since

$$h = s \, \sin(\theta),$$

then

$$v^2 - 2gs \, \sin(\theta) = 2\mu mgs \, \cos(\theta),$$

or

$$v = \sqrt{2gs \, [\mu \cos(\theta) + \sin(\theta)]}.$$

In auto accident investigation work it is often the case that θ is "small," and so, using the approximations $\sin(\theta) \approx \theta$ (with θ in radians)[7] and $\cos(\theta) \approx 1$, we have the speed v at the start of the skid as

$$v = \sqrt{2gs \, [\mu + \theta]},$$

a formula commonly used by police officers investigating auto accidents. To use it we obviously need the value of μ, which is easily found by performing a test skid using a car similar to the one involved in the incident, at the location of the incident. So, suppose the test vehicle is driven up to the posted speed of 25 mph and then put into a skid, generating skid marks 46.5 feet long. From our formula for v, we have

$$\mu + \theta = \frac{v^2}{2gd} = \frac{\left[\left(\frac{25}{60}\right) \times 88\right]^2}{2 \times 32.2 \times 46.5} = 0.45,$$

in which I've used the conversion (worth memorizing!) of 60 mph = 88 feet/second (fps). Notice that this result is *not* the value of μ itself (unless $\theta = 0$) but a value that includes the effect of the road slope. In particular, for the preface problem, in which the skid marks going into an uphill skid were 106 feet long, the speed entering the skid must have been

$$v = \sqrt{2 \times 32.2 \times 106 \times 0.45} = 55.4 \text{ fps} \approx 38 \text{ mph}.$$

The driver was clearly speeding.

Now, what about the second part of the preface problem: how does the calculation change if all is as before except that now the 106-foot skid occurs on an 8% *downhill* slope? That is, if the speed we just calculated is called v_u (for uphill), what is v_d (for downhill)? For an uphill calculation we have, as we've already worked out for $\theta > 0$,

$$v_u^2 = 2gs[\mu + \theta].$$

For the downhill case we simply substitute $-\theta$ for θ, and so

$$v_d^2 = 2gs[\mu - \theta].$$

Thus,

$$v_u^2 - v_d^2 = (2gs\mu + 2g\theta s) - (2gs\mu - 2g\theta s) = 4g\theta s,$$

or, as for an 8% slope, we have $\theta = 0.08$,

$$\begin{aligned} v_d^2 &= v_u^2 - 4g\theta s \\ &= \{(55.4)^2 - 4 \times 32.2 \times 0.08 \times 106\} \text{ feet-squared/seconds-squared} \\ &= 1{,}977 \text{ feet-squared/seconds-squared.} \end{aligned}$$

Thus,

$$v_d = 44.5 \text{ fps} = 30.3 \text{ mph.}$$

This means that the driver was technically still speeding in a 25 mph zone but, now, much less flagrantly than in the uphill case.

Notes

1. One horsepower is equal to 746 watts. If you're wondering where this curious number originated, let me just say it's all a historical accident dating from the earliest days of the steam engine. The history *is* interesting, but it's *physics* we're after here, and I'll say no more.

2. See Geoffrey T. Fox, "On the Physics of Drag Racing," *American Journal of Physics*, March 1973, pp. 311–313. Fox was a professor of physics at the University of Santa Clara (California) when he wrote this paper, but he later left academia to start Fox Racing USA.

3. Fox's paper gives the numerical coefficient as 270 rather than 281, but he simply states this value without showing any numerical work. I suspect the 270 is a typo.

4. John W. Burgeson, "A Problem in Free Space Dynamics," *American Journal of Physics*, April 1956, p. 288.

5. Since the frictional retarding force *decreases* when skidding occurs, it is clearly desirable *not to skid* in an emergency stop. That is, it is desirable to not lock the wheels, and this is the idea behind ABS (antilock braking system), found on many cars. Owners of cars equipped with ABS are generally given a discount on their auto insurance premium because of the inherent increase in driving safety.

6. A *skidding* tire becomes hot, due to friction. So some of the kinetic energy of the skidding car goes into heating the tires. As a first approximation, I'll ignore this effect.

7. For $\sin(\theta) = 0.08$, $\theta = 4.6°$, an angle most people (I think) would consider "small."

5. Merry-Go-Round Physics and the Tides

So David prevailed over the Philistine with a sling and with a stone.

— *Goliath met his biblical end because David understood centripetal acceleration (Samuel I:50)*

When you tie a rock to one end of a rope and then, grasping the other end, swing the rock in a circle around your head, you are observing *centripetal force* in action. You personally experience that force when riding on a merry-go-round, and the entire Earth experiences it as it orbits the Sun. In the first example, the force *on the rock* is the tension in the rope (the force *you* feel, equal in magnitude but in the opposite direction, is what is commonly called the *centrifugal* force); "in the second example the centripetal force is the force your body exerts on the merry-go-round to keep you from flying off the spinning platform, and in the last example the centripetal force is due to gravity.

A common misconception is that if you let go of the rock (or of the merry-go-round) the rock (or you) will zip directly (radially) away from the center of rotation. This is not so; the resulting motion will be *tangent* to the circular path. When David took on Goliath with his slingshot, it was important that he had the physics right!

If a mass m moves at speed v in a circle of radius R, then it is continually being forced to bend away from a straight-line path into a curved path. That force comes from an *inward* acceleration toward the center of the circle, an acceleration with value $\frac{v^2}{R}$. From Newton's

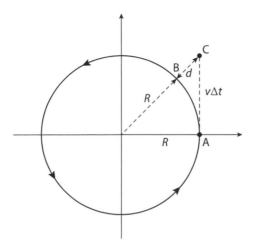

Figure 5.1. Centripetal force in action

second law the *inward*-directed force F (the word *centripetal* means "center seeking") has *magnitude*[1]

$$F = m\frac{v^2}{R}.$$

Here's a simple derivation of the centripetal acceleration of circular motion.

In Figure 5.1 we have a mass m moving at the constant speed v along a circular path of radius R. The *velocity* of the mass, which like force is a *vector*, has *magnitude* v, which is called the *speed* (speed is a scalar). While the speed is a constant, the *direction* of the velocity is obviously always changing. This change in direction is caused by a force. That's because at time $t = 0$ the mass is at A in Figure 5.1, which, with no loss in generality, we can take to be on the horizontal axis of our coordinate system. Then, a short time Δt later the mass is at B. *If there was no force acting on the mass, however, it would instead have been at* C, a vertical distance $v\Delta t$ upward from A. But the mass m isn't at C because it was acted on by a force, the force that we are now going to calculate.

Point C is shown in Figure 5.1 to be distance d from B, and so some force must have "pulled" the mass inward through distance d to keep the moving mass on the circular path. If that force caused the constant acceleration a, then we must have

$$d = \frac{1}{2}a(\Delta t)^2.$$

From the Pythagorean theorem you can see that

$$R^2 + (v\Delta t)^2 = (R+d)^2 = R^2 + 2Rd + d^2,$$

or

$$v^2(\Delta t)^2 = (2R+d)d = \left[2R + \frac{1}{2}a(\Delta t)^2\right]\frac{1}{2}a(\Delta t)^2.$$

Canceling $(\Delta t)^2$ on both sides gives

$$v^2 = \left[2R + \frac{1}{2}a(\Delta t)^2\right]\frac{1}{2}a = Ra + \frac{1}{4}a(\Delta t)^2,$$

and so, as we let $\Delta t \to 0$,

$$v^2 = Ra.$$

That is, the centripetal acceleration is

$$a = \frac{v^2}{R}.$$

From $F = ma$ we then have the force equation discussed in the opening of this chapter.

One of the great successes of Newton's studies was his discovery that the gravitational field *outside* a massive, spherically symmetric body is exactly what it would be if the massive body were a *point* mass.[2] So, when calculating the orbit of the Earth around the Sun we can replace both the Sun and the Earth with point masses, because the orbit is outside the Sun (obvious, I think!). The "orbit" is the path of the Earth's center. That is, if the masses of the Sun and the Earth are M and m, respectively, then the gravitational force (a centripetal force) on the Earth is given by Newton's famous *inverse-square law*:

$$G \frac{Mm}{r^2}$$

where r is the radius of the Earth's orbit (measured from the center of the Sun to the center of the Earth), and G is the *universal gravitational constant*.[3]

Here's an interesting question for you using Newton's law of gravity. Which do you think exerts the greater gravitational force on Earth, the Sun or the Moon? The Sun is a lot bigger than the Moon, but it is also a lot farther from the Earth than is the Moon. These two parameters, mass and distance, are in opposition, and so it isn't immediately obvious which one dominates. We can calculate the answer as follows, using the following numerical values:

mass of the Sun $= 2 \times 10^{30}$ kilograms $= M_s$
mass of the Moon $= 7.35 \times 10^{22}$ kilograms $= M_m$
Earth/Sun separation $= 93 \times 10^6$ miles $= R_s$
Earth/Moon separation $= 2.39 \times 10^5$ miles $= R_m$

So, the ratio of the gravitational force on Earth due to the Sun to the gravitational force on Earth due to the Moon is (m is the mass of the Earth, which cancels, and so we don't need to know it)

$$\frac{G \frac{M_s m}{R_s^2}}{G \frac{M_m m}{R_m^2}} = \frac{M_s}{M_m} \left(\frac{R_m}{R_s} \right)^2 = \frac{2 \times 10^{30}}{7.35 \times 10^{22}} \left(\frac{2.39 \times 10^5}{93 \times 10^6} \right)^2 = 180.$$

The Sun's gravitational force on Earth is 180 times greater than is the Moon's.

We can now derive one of the fundamental results of orbital physics, a result due to the German astronomer Johannes Kepler (1571–1630), who deduced it in 1619 from tedious experimental observation of the motions of the visible planets, decades before Newton was born. Using Newton's gravitational force law, however, we can *derive* Kepler's result while sitting comfortably in front of a cozy fireplace without once looking upward to the heavens.

Suppose the mass m orbits the mass M once in time T at a fixed distance r at speed v. Newton's force law, combined with centripetal acceleration, says

$$G\frac{Mm}{r^2} = m\frac{v^2}{r},$$

and so

$$GM = v^2 r.$$

Since

$$T = \frac{2\pi r}{v},$$

then

$$v = \frac{2\pi r}{T},$$

and so[4]

$$GM = \frac{4\pi^2 r^2}{T^2}r = \frac{4\pi^2}{T^2}r^3,$$

or

$$\frac{r^3}{T^2} = \frac{GM}{4\pi^2} = \text{constant for a given } M.$$

This is Kepler's so-called third law of planetary motion, where M is the mass of the Sun. Notice that m, the mass of the orbiting object, doesn't appear in the equation, and so the constant is the same for all the planets orbiting the Sun.[5]

Now, finally, we come to the question of the ocean tides. To start, forget the Moon and concentrate on the Earth orbiting the Sun, as shown in Figure 5.2. There the massive Sun is drawn as a point mass and the relatively tiny Earth as an extended object (the figure is NOT

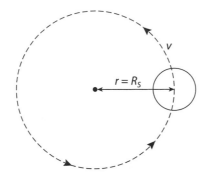

Figure 5.2. The Earth orbiting the Sun

to scale!). Imagine Earth covered with water (as it mostly is). The *center* of the Earth is distance R_s from the center of the Sun, and that is the value for r we use in Newton's gravitational force law. As we've already written in the discussion on Kepler's third law, the orbital speed of the Earth is given by

$$v = \frac{2\pi R_s}{T},$$

and, of course, this is the orbiting speed of not just the center of the Earth but of *all* the Earth (this is an observational fact, because otherwise we'd see the Earth tearing apart!).

Now, the water on the side of the Earth closest to the Sun is *not* at distance R_s from the center of the Sun but, rather, is at distance $R_s - R$, where R is the radius of the Earth. So, the gravitational force on that side of the Earth is *greater* than is required for the water to orbit at speed v, and this extra attraction forms a bulge of water toward the Sun. This bulge is one that most people find intuitively obvious. But what they don't find obvious is that there is a second bulge directly opposite the first bulge, on the other side of the Earth, the side *farthest* from the Sun. The explanation for that second bulge, however, is exactly the same: centripetal acceleration. That is, the water on the side of the Earth facing away from the Sun is not at distance R_s from the center of the Sun but, rather, is at distance $R_s + R$. So, the gravitational force on that side of the Earth is *less* than is required for the water to orbit at speed v, and this reduced attraction allows a bulge of water to form pointing *away* the Sun.

These two bulges are fixed along a line joining the centers of the Sun and the Earth, but since the Earth rotates on an axis tilted about 23° from the perpendicular to the orbital plane, the two bulges move around the Earth (more accurately, the Earth rotates *under* the two bulges once every 24 hours), and an encounter—when we see one of the bulges every 12 hours—is what we call the *solar high tide*. Since the Earth and the Moon are also revolving around each other (about a common center of mass[6]), there are lunar tides as well. Recall that we earlier calculated the Sun to be 180 times stronger gravitationally on Earth than is the Moon. That might naively lead you to expect the lunar high tides to be insignificant compared with the solar high tides. But that's not so, and in fact, it's just the opposite. *Lunar* tides are the larger tides. Why?

It's because the Sun is so much farther from the Earth than is the Moon. I've stated that the tides are caused by the *variation* of the gravitational force at the near and far sides of the Earth, compared with the gravitational force at the Earth's center. For the Sun, this *variation* is smaller than is the variation for the Moon. We can calculate the gravity variations on a mass m as follows. For the Sun, the

$$\text{gravity at the near side of Earth} = G\frac{M_s m}{(R_s - R)^2},$$

and the

$$\text{gravity at the far side of Earth} = G\frac{M_s m}{(R_s + R)^2},$$

and so, over the diameter of the Earth, the variation on Earth of the Sun's gravity is

$$G\frac{M_s m}{(R_s - R)^2} - G\frac{M_s m}{(R_s + R)^2} = GM_s m\left[\frac{1}{(R_s - R)^2} - \frac{1}{(R_s + R)^2}\right],$$

which reduces (with the approximation $R_s \gg R$) to

$$4GM_s m\frac{R}{R_s^3}.$$

Notice that this variation varies inversely with the *cube* of the distance. Similarly, the variation from the near side to the far side of the Earth of the Moon's gravity is

$$4G\,M_m\,m\,\frac{R}{R_m^3}.$$

So,

$$\frac{\text{Moon variation}}{\text{Sun variation}} = \frac{M_m}{M_s}\left(\frac{R_s}{R_m}\right)^3 = \frac{7.35 \times 10^{22}}{2 \times 10^{30}}\left(\frac{93 \times 10^6}{2.39 \times 10^5}\right)^3 = 2.16.$$

The lunar tides are more than twice as large as the solar tides, even though the Sun is far more massive than the Moon. That size advantage of the Sun is overcome by its extreme distance from the Earth; over the diameter of the Earth, the Sun's gravity hardly varies as we go from 93 million minus 4,000 miles to 93 million plus 4,000 miles.

Tidal forces due to gravitational variation across the extent of a body have had a spectacular result in our Solar System, in addition to the ocean tides on Earth. Such forces are the cause (or at least are *suspected* to be the cause) of the beautiful rings of Saturn. Long ago, it is believed, a moon of Saturn got too close to the huge planet and was literally pulled apart by the planet's tidal force on it, with the resulting multitude of fragments forming what we see today as the rings.[7]

Finally, one last technical comment on the fact that there are *two* lunar tidal bulges: they are the result of the *motion* of the Earth-Moon system. If the Earth and the Moon were fixed in space, with the Earth's rotation about an axis its only motion, then there would be just *one* high tide on Earth each 24 hours, directly below the Moon. It is the "orbital" motion of the Earth about the Earth-Moon center of mass that gives rise to the second, far-side ocean tidal bulge on Earth.

Ancient Chinese writers thought that the oceans were the blood of the Earth, that the tides reflected the beating of the Earth's pulse, and that the tides were caused by the Earth breathing. That is all very romantic stuff, inspired by the equally ancient mythological idea of Gaia, that the Earth is a living organism. But this is a *physics* book, not one of poetry, and so I repeat: *gravity* is what is behind the tides.

Notes

1. To distinguish the *magnitude* of a vector (force) from the vector itself, textbook authors use one of various typesetting formats, for example, $\vec{F}$ for the force vector and F for the magnitude (that is, $F = |\vec{F}|$), or boldface for the vector (and so $F = |\mathbf{F}|$).

2. This is one of two results Newton published in his 1687 *Principia* that are, together, called his *superb theorems*. (The other is that the gravitational force on a point mass inside a hollow, spherical shell of uniformly distributed matter, *no matter where inside the shell the point mass is located*, is zero.) You can find modern, calculus-based derivations of both theorems (Newton used *very* involved geometric arguments) in my book *Mrs. Perkins's Electric Quilt*, Princeton University Press, 2009, pp. 140–147.

3. The story of G and of its connection to the famous *Cavendish experiment* (an experiment so delicate it wasn't performed until 1798, seventy-one years after Newton's death) is told in *Mrs. Perkins's*, pp. 136–140. The value of G is $6.67 \times 10^{-11} \frac{\text{m}^3}{\text{kg·s}^2}$. As of publication of this book (2016) the value of G is still known to only three significant digits, far fewer than we know most other physical constants; see Clive Speake and Terry Quinn, "The Search for Newton's Constant," *Physics Today*, July 2014, pp. 27–33, and the end of Chapter 22.

4. At a distance of 93,000,000 miles from the Sun, and with an orbital period of 365 days, the orbital speed of the Earth around the Sun is slightly more than 18 miles/second.

5. This is strictly true only if $M \gg m$, which is the case for the Solar System. You can read about all three of Kepler's laws (including a derivation of the nature of the dependency of the "constant in the third law on m), in *Mrs. Perkins's*, pp. 170–185.

6. Because the Earth is so much more massive than the Moon, their center of mass is actually *inside* the Earth, more than a thousand miles below the planet's surface. For details, see *Mrs. Perkins's*, pp. 175–178.

7. A wonderfully gruesome science fiction story, based on the tidal force a massive body can exert on a physically "small" mass (where *small* in the tale is the distance between the head and the feet of a human space traveler) that gets too close to a superdense star, is the 1966 classic "Neutron Star" by Larry Niven. The result is science fiction's version of the medieval torture rack.

6. Energy from Moving Water

[The] energy of the tides is continuously being dissipated at
a rate whose order of magnitude is a billion horsepower!"
— *Edward P. Clancy,* The Tides: Pulse of the Earth *(1968)*

In Chapter 3 we found that there is substantial energy in moving air. How about moving water? For example, how much energy is in the ocean tides of the entire planet? The answer is *a lot* (Clancy is actually on the *low* side with his estimate), and calculation of that energy requires only the application of simple physics. We start with the origin of the tides—the Moon (and the Sun to a lesser extent)—as discussed in the previous chapter. There we saw how gravity and centripetal acceleration creates two tidal bulges, one directly below the Moon, and the other on the far side of the Earth opposite the first bulge. The two tidal bulges appear to move around the Earth as it rotates on its polar axis, and so we see a "high tide" every 12 hours.

But here's a new twist on matters that we didn't discuss in Chapter 5. Because of frictional forces, the two bulges are not directly in-line with the centers of the Earth and the Moon, as in Figure 6.1 but, rather, are actually slightly offset, as shown in Figure 6.2. The reason for that offset is that there isn't perfect elasticity or fluidity in the Earth's solid and liquid surface components, respectively. Because of these frictional forces the Earth's surface does not respond instantly to forces, and so the rotation of the Earth carries the tidal bulges forward.

The Moon's gravitational pull on the two tidal bulges produces a net counter-rotational torque[1] that tends to reduce the Earth's rotational speed. The Moon's pull on the far-side bulge tends to increase that

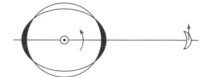

Figure 6.1. The tidal bulges with no friction

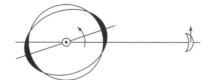

Figure 6.2. The tidal bulges with friction

speed, but the pull on the near-side bulge is greater, and that pull tends to decrease the rotational speed. The net result is that the Earth's rotation is slowing (that is, the length of a day is continually increasing). But this is occurring *very* slowly. Atomic clocks have shown that the length of a day is increasing at the rate of about 2 *milli*seconds per *century*! That is, the length of a day 100 years ago was just 0.002 seconds shorter than was yesterday, the length of a day 200 years ago was just 0.004 seconds shorter than was yesterday, and so on.

An entirely different way of arriving at the length of ancient days comes from marine biology. Examinations of growth patterns in the skeletal structure of fossil coral reefs from the Middle Devonian period (375 million years ago), patterns sensitive to daily and seasonal variations in the environment, indicate there were about 400 days to the year then. Since the length of a year is invariant, due only to Earth's orbital mechanics, then the length of a Middle Devonian day must have been $\frac{365}{400}$ (24) hours = 21.9 hours. So, 3,750,000 centuries ago the day was 2.1 hours shorter. That is, *per century*, the length of a day changed by $\frac{2.1 \times 3,600}{3,750,000}$ seconds = 0.002 seconds.

You may wonder how a steady increase in the length of a day, amounting to just 2×10^{-3} seconds after a century, can be important, but you must understand that the effect is cumulative. If, for example, we assume this rate of increase has been in effect for the last 2,000 years

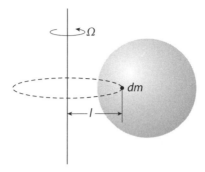

Figure 6.3. A rotating mass built up from differential mass elements *dm*

(20 centuries), then the day Julius Caesar was assassinated (44 BC) was shorter in duration, compared with yesterday, by $2 \times 10^{-3} \times 20 = 40$ milliseconds. Since this result reflects the steady decrease in the length of a day going backward in time, then the *average* change in the duration of *each one* of the days during the past 2,000 years was 20×10^{-3} seconds. So, the *total accumulated* shift in the timing of an event 2,000 years ago would be

$$20 \times 10^{-3} \frac{\text{seconds}}{\text{day}} \times 2{,}000 \, \text{years} \times 365 \frac{\text{days}}{\text{year}} = 14{,}600 \, \text{seconds},$$

that is, 4 *hours!*[2]

The Earth is sufficiently massive that shifting time by four hours over 2,000 years requires a stupendously enormous amount of energy, and it is this energy that we'll calculate in this chapter. To make the calculation, we'll first have to explore how to calculate the *rotational* kinetic energy of the Earth, so we'll begin with that.[3]

Imagine an extended three-dimensional body of mass M and volume V revolving at a constant angular rate about some *axis of rotation*, as shown in Figure 6.3. The angular rotation rate is Ω radians/second, which means that if T is the time for one complete rotation (in seconds), then

$$\Omega T = 2\pi.$$

As shown in Figure 6.3, the massive body M is constructed from differential mass elements dm, each a variable distance l from the axis of rotation. Since each element is moving because of the rotation, then each element has differential kinetic energy dE, given by

$$dE = \frac{1}{2}(dm)v^2,$$

where the speed of each element is

$$v = \Omega l.$$

So,

$$dE = \frac{1}{2}\Omega^2 l^2 dm,$$

and if we integrate dE over the entire spatial extent of the body, we'll get the total energy of rotation about the given axis:

$$E = \iiint_V dE = \frac{1}{2}\Omega^2 \iiint_V l^2 dm = \frac{1}{2}\Omega^2 I.$$

We can take Ω^2 outside the triple integral because Ω is a constant, but we must leave l^2 inside because the distance of each dm mass element from the axis of rotation varies (in general). The rightmost triple integral is I, the *moment of inertia* of the body about the given axis of rotation. *Note*: For the examples in this book, the massive bodies we'll consider will have lots of symmetry, and we'll not actually have to do any *triple* integrals.

As a simple example of such a calculation, consider a right circular cylinder of radius R, height h, and constant mass density ρ, with its long axis as the y-axis, taken as the axis of rotation, as shown in Figure 6.4. Imagine that the cylinder is made from layers of cylindrical shells, of radius x, where $0 \le x \le R$, with a wall thickness of dx. That is, the cylinder is constructed from *hollow shells* with an inner radius of x and an outer radius of $x + dx$. The differential mass element dm of the solid cylinder is the mass of a shell, given by $dm = \rho 2\pi x h dx$. For each such shell, the *differential* moment of inertia around the y-axis is

$$dI_{\text{shell}} = x^2(\rho 2\pi x h dx) = \rho 2\pi h x^3 dx,$$

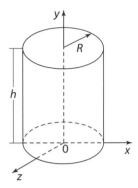

Figure 6.4. A solid cylinder, rotating about the y-axis

where we must write the left-hand side as a *differential* moment because of the dx on the right-hand side of the equation.

To calculate the moment of inertia for a solid cylinder, we think of the *solid* cylinder as composed of infinitely many layers of cylindrical, thin-walled *shells* of increasing radius. Mathematically this means we integrate $d I_{\text{shell}}$ over all x from 0 to R. Thus,

$$I_{\text{solid}} = \int_0^R d I_{\text{shell}} = \int_0^R \rho 2\pi h x^3 \, dx = \rho 2\pi h \left(\frac{x^4}{4} \right) |_0^R = \frac{\rho \pi h}{2} R^4.$$

The total mass of the solid cylinder is

$$M_{\text{solid}} = \pi R^2 \rho h,$$

and so

$$I_{\text{solid}} = \frac{1}{2} M_{\text{solid}} R^2.$$

For a *thin*-walled cylindrical shell of radius R rotating about its long axis, *all* the mass is located at the same distance from the axis, and so, by inspection, we have

$$I_{\text{shell}} = M_{\text{shell}} R^2.$$

Now, to calculate the moment of inertia for the Earth, we need to evaluate the triple integral $\iiint_V l^2 dm$ for the case where the axis of rotation is a diameter of a solid sphere. You can find a brute-force evaluation of this integral for a sphere in most books on freshman calculus, but let me show you a *clever* way to do it. We'll do it in two steps, first finding the triple integral for a spherical *shell*, that is, a hollow sphere with a very thin surface (think of a balloon). Then, just as we did for a solid cylinder, we'll use our spherical shell result to extend the shell to a solid sphere.

Step 1: We start with a thin-skinned-surface shell of radius a, a shell thickness of da, and a constant mass density of ρ. All the dm elements *are* the surface, and so we write

$$dm = \rho \, dS \, da,$$

where dS is a differential-area patch on the surface. That means if we integrate dS over the entire surface of the shell, we'll get the total surface area of the spherical shell:

$$\iint_S dS = S = 4\pi a^2.$$

Now, imagine that the axis of rotation is the x-axis, which means that the distance of each element dm from the axis of rotation depends only on its y- and z-coordinates, because *on* the surface we have $x^2 + y^2 + z^2 = a^2$, and so x is determined once y and z are given. That is,

$$l^2 = y^2 + z^2.$$

Since all the mass of the shell is on the surface and there is none in the interior, the triple integral reduces to a double integral over the surface, and we get

$$dI_x = \iint_S \left(y^2 + z^2\right) \rho \, dS \, da = \rho \, da \iint_S \left(y^2 + z^2\right) dS.$$

In the same way, if the axis of rotation was the y-axis, we would have

$$l^2 = x^2 + z^2,$$

and if the axis of rotation was the z-axis we would have

$$l^2 = x^2 + y^2.$$

So,

$$dI_y = \rho\, da \iint_S \left(x^2 + z^2\right) dS,$$

and

$$dI_z = \rho\, da \iint_S \left(x^2 + y^2\right) dS.$$

Here's the clever observation I promised you: by the symmetry of a sphere,

$$dI_x = dI_y = dI_z = dI_{\text{shell}}.$$

Pretty obvious, *once it's pointed out*, and it's all we need to finish our calculations. From our earlier expressions, we have

$$dI_x + dI_y + dI_z = 3dI_{\text{shell}} = \rho\, da \iint_S \left(y^2 + z^2\right) dS + \rho\, da \iint_S \left(x^2 + z^2\right) dS$$

$$+ \rho\, da \iint_S \left(x^2 + y^2\right) dS = 2\rho\, da \iint_S \left(x^2 + y^2 + z^2\right) dS.$$

Now, as stated earlier, on the surface of the spherical shell (where all the shell mass is located) we have

$$x^2 + y^2 + z^2 = a^2,$$

and so

$$3dI_{\text{shell}} = 2\rho\, da \iint_S a^2 dS = 2\rho a^2 da \iint_S dS = 2\rho a^2 da\, (4\pi a^2) = 8\pi\rho a^4 da.$$

Thus, the differential moment of inertia of a spherical shell of radius a and shell thickness da (around *any* diameter axis of rotation) is

$$dI_{\text{shell}} = \frac{8}{3}\pi\rho a^4 da.$$

Step 2: To find the moment of inertia for a *solid* sphere of radius R, we imagine it to be like an onion, that is, composed of *infinitely many layers of spherical shells* of increasing radius. Mathematically, this means we integrate our shell result over $0 \leq a \leq R$. Thus, for a solid sphere with constant mass density ρ,

$$I_{\text{solid}} = \int_0^R dI_{\text{shell}} = \frac{8}{3}\pi\rho \int_0^R a^4 da = \frac{8}{3}\pi\rho \left(\frac{a^5}{5}\right) \Big|_0^R = \frac{8\pi R^5}{15}\rho.$$

The mass of the solid sphere is

$$M_{\text{solid}} = \frac{4}{3}\pi R^3 \rho,$$

and so, for a constant-density sphere,

$$I_{\text{solid}} = \frac{2}{5} M_{\text{solid}} R^2 = 0.4 M_{\text{solid}} R^2.$$

The Earth, however, is *not* a constant-density sphere, as its central regions are much denser than the regions nearer the surface.[4] Consequently, the Earth's moment of inertia is given by a coefficient *smaller*[5] than 0.4; specifically,

$$I_{\text{Earth}} = 0.3444 M_{\text{solid}} R_{\text{Earth}}^2.$$

Now, we are all set to calculate the power of the ocean tides (where from now on I'll drop the "Earth" subscript). The rotational kinetic energy of the Earth is

$$E = \frac{1}{2}\Omega^2 I,$$

where

$$\Omega = \frac{2\pi}{T},$$

where T is the rotational period of the Earth (the length of a day), and so

$$E = \frac{1}{2}\frac{4\pi^2}{T^2}(0.3444)MR^2 = 0.6888M\frac{\pi^2 R^2}{T^2} = \frac{C}{T^2}.$$

(I've introduced the constant $C = 0.6888 M\pi^2 R^2$, with units of kilograms · meters-squared, to help keep the math transparent.) Notice that as a result, E has units of $\frac{\text{kilograms·meters-squared}}{\text{seconds squared}}$, and you should check to be sure these are indeed the units of energy to be sure all is consistent. Recall that energy is force times distance (see note 2 in Chapter 3), and that force is mass times acceleration, and so energy has units of mass times acceleration times distance, or

$$\text{kilograms} \times \frac{\text{meters}}{\text{seconds squared}} \times \text{meters} = \frac{\text{kilograms · meters-squared}}{\text{seconds squared}},$$

just as we have found. The unit of energy is called the joule (see note 4 in Chapter 3), and so

$$1\,\text{joule} = 1\frac{\text{kilogram · meters-squared}}{\text{seconds squared}}.$$

Now, if $E + \Delta E$ is the rotational kinetic energy of the Earth, when the time for one rotation of the Earth has increased from T to $T + \Delta T$, then

$$E + \Delta E = \frac{C}{(T + \Delta T)^2},$$

and so

$$\Delta E = \frac{C}{(T + \Delta T)^2} - E = \frac{C}{(T + \Delta T)^2} - \frac{C}{T^2} = C\left[\frac{1}{(T + \Delta T)^2} - \frac{1}{T^2}\right]$$

$$= C\left[\frac{T^2 - (T + \Delta T)^2}{T^2 (T + \Delta T)^2}\right] = C\frac{T^2 - T^2 - 2T\Delta T - (\Delta T)^2}{T^2 \left[T^2 + 2T\Delta T + (\Delta T)^2\right]},$$

or, assuming $T \gg \Delta T$,

$$\Delta E \approx -C\frac{2T\Delta T}{T^4} = -2C\frac{\Delta T}{T^3}.$$

Taking $T = 86,400$ seconds and $\Delta T = 2 \times 10^{-3}$ seconds (thus confirming our earlier assumption of $T \gg \Delta T$), we have

$$\Delta E \approx -2\,(0.6888)\,M\pi^2 R^2 \frac{2 \times 10^{-3}}{\left(8.64 \times 10^4\right)^3}.$$

Using $M = 5.98 \times 10^{24}$ kilograms for the mass of the Earth, and $R = 6.38 \times 10^6$ meters for the radius of the Earth, we have the change in the Earth's rotational kinetic energy over a 100-year interval (100 years because we took $\Delta T = 2$ milliseconds) as

$$\Delta E \approx -2\,(0.6888)\,\left(5.98 \times \text{kilograms}\right)\pi^2$$
$$\times \left(6.38 \times 10^6 \text{ meters}\right)^2 \frac{2 \times 10^{-3} \text{ seconds}}{\left(8.64 \times 10^4 \text{ seconds}\right)^3}$$
$$= 10.26 \times 10^{21} \text{ joules}.$$

Dividing this energy by the number of seconds in 100 years (3.15×10^9), we get a *power* of 3,260 gigawatts. Since 1 horsepower = 746 watts, the power of the ocean tides is $\frac{3,260 \times 10^9}{746}$ horsepower = 4.37 billion horsepower.

That's a pretty big number, and it isn't surprising that long before our modern times, people thought about accessing some of that power. One interesting idea that looks good at first glance (but actually isn't) was described as follows:

> I saw some years ago a suggestion that the rise and fall of old hulks on the tide would afford serviceable power. If we picture to ourselves the immense weight of a large ship, we may be deluded for a moment into agreement with this project, but numerical calculation soon shows its futility. The tide takes about six hours to rise from low water to high water, and the same period to fall again. Let us suppose that the water rises ten feet, and that a hulk of 10,000 tons displacement is floating on it; then it is easy to show that only twenty horse-power will be developed. . . . I am glad to say that the projector of this scheme gave it up when its relative insignificance was pointed out to him.[6]

The "only twenty-horsepower" is actually an *over*estimate. That's be-cause to raise (or lower) 10,000 tons (20,000,000 pounds) 10 feet involves 200,000,000 foot-pounds of energy. Since this energy is developed in 6 hours (21,600 seconds), the *power* is

$$\frac{200,000,000 \text{ foot-pounds}}{21,600 \text{ seconds}} = 9,259 \frac{\text{foot-pounds}}{\text{second}}.$$

Since 1 horsepower $= 550 \frac{\text{foot-pounds}}{\text{second}}$, we have a power level of

$$\frac{9,259}{550} \text{horsepower} = 16.8 \text{ horsepower},$$

a value that makes Darwin's point even stronger.

I'll stop here for now, and give you a chance to ruminate on all our discussions on rotational physics. We'll return to these ideas in Chapter 10, and there we'll extend them a bit to answer some additional interesting physics questions left over from earlier in the book (why the Moon is receding from the Earth, how a falling chimney buckles, and how fast a cylinder can roll down an inclined plane.)

Notes

1. The torque on the Earth due to the Moon's gravity acting on a tidal bulge is the product of a force with the length of a lever arm. (Think of the torque you apply to a nut with a wrench when on your back under the kitchen sink! The units of torque are foot-pounds in the English system, and newton-meters in the metric system. While the units of torque and energy are the same, they are very different concepts.) The force in the Earth/Moon-torque system is the component of the gravitational force on the bulge that is *perpendicular* to the line joining the Earth's center to the bulge, and the torque lever arm is that line (its length is of course the radius of the Earth).

2. Early (that is, preatomic clock) researchers tried to use this idea in reverse, to determine the rate of slowing, by comparing the reported timing of ancient eclipses with the timing predicted by Newton's theory of gravity under the assumption of a *constant* length to the day. Those attempts were not very successful—see Walter H. Munk and Gordon J. F. MacDonald, *The Rotation of the Earth*, Cambridge University Press, 1960, pp. 186–191.

3. Earth has *translational* kinetic energy because it orbits the Sun, but even if that orbital motion ceased, the planet would still have rotational kinetic energy, because it is spinning on its polar axis. The two kinetic energies add to give the total kinetic energy of the Earth.

4. For details on the density of the inner regions of the Earth, see *Mrs. Perkins's*, pp. 191–200. At its center, the Earth's density is about 13 times that of water, while near the surface it is around 3 times that of water.

5. *Smaller* because a relatively greater fraction of the Earth's mass is located nearer the Earth's axis of rotation than would be in a constant-density sphere.

6. From Sir George Darwin, *The Tides* (pp. 73–74), originally published in 1898 and reprinted by W. H. Freeman in 1962. Darwin displayed a dry sense of humor when, in his very next sentence, he wrote: "It is the only instance of which I ever heard where an inventor was deterred by the impracticability of his plan." Darwin was, in fact, not at all enthusiastic about the extraction of energy from the tides, preferring *rivers* as water-based energy sources.

7. Vectors and Bad Hair Days

It is impossible to travel faster than the speed of light, and
certainly not desirable, as one's hat keeps blowing off.
— *Woody Allen*

At the local shopping mall that my wife and I frequent, I've noticed
there is often a stiff, steady wind blowing over the vast flat asphalt
parking lot we walk across to get to a mall entrance. There are
a number of available entrances, and so I've taken to picking that
entrance that lets the wind that day come as close as possible to blowing
at a right angle (left to right) to my path. That's because then the
wind blows my hair (what's left of it) flat rather than all over my head.
This desire to maintain sartorial elegance in the food court (where I'm
writing this) motivates the following pretty (but simple) little problem
in vector physics.[1]

Given that I initially am walking with the wind, through what angle
should I turn so that the wind appears (to me) to be blowing at a right
angle to my path? Surprisingly (perhaps) to many, the answer is *not* 90°.

Let's write $\vec{v}$ as my velocity vector relative to the parking lot (my so-
called ground velocity) and $\vec{w}$ as the velocity vector (also relative to the
ground) of the wind. Then, my velocity vector *relative to the wind* is $\vec{v}'$,
where

$$\vec{v}' = \vec{v} - \vec{w}.$$

This should make immediate physical sense in the two special cases:
(1) I'm walking with the wind and thus $\vec{v}$ and $\vec{w}$ are parallel (and so
$v' = v - w)^2$, and (2) I'm walking against the wind and thus $\vec{v}$ and $\vec{w}$

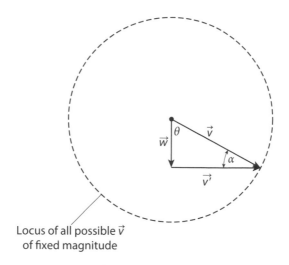

Locus of all possible $\vec{v}$
of fixed magnitude

Figure 7.1. The case of $w < v$

are antiparallel (and so $v' = v + w$). Using vectors allows a simple way to express all the other possibilities of how I walk with respect to the wind in a single expression. Rewriting our vector equation, we have

$$\vec{v} = \vec{v}' + \vec{w}.$$

Now, if my velocity relative to the wind is $\vec{v}'$, then $-\vec{v}'$ is the wind's velocity relative to me, which is what my hair responds to as I walk across the parking lot. In Figure 7.1 I drew the vector $\vec{w}$ pointing straight down, something we can always do because $\vec{w}$ is a given, fixed vector and we'll just *define* the wind's direction to be the *down* direction (draw $\vec{w}$ in whatever direction you like and then rotate the paper until $\vec{w}$ points down!). The vector $\vec{v}'$ is then the vector that has to be added to $\vec{w}$ to give $\vec{v}$. Remember, $\vec{w}$ is a given vector, while $\vec{v}$ is a vector of our choice. Once we've picked $\vec{v}$ for the stroll to the mall, it and the given $\vec{w}$ determine $\vec{v}'$. Now, we want to choose $\vec{v}$ so that $\vec{v}'$ (actually, $-\vec{v}'$) is perpendicular to $\vec{v}$ (so that $\alpha = 90°$). In Figure 7.1 I've assumed $w < v$ (that the wind is blowing slower than I walk), and you see, as we let $\vec{v}$ rotate around a full circle (keeping its *magnitude* fixed), that there is never a choice for $\vec{v}$ (there is no turning angle θ) that gives a perpendicular $\vec{v}'$! I don't think this is obvious a priori.

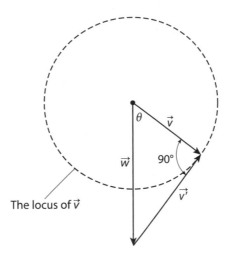

Figure 7.2. The case of $w > v$

The situation is different in Figure 7.2, where the assumption is now that $w > v$ (the wind is blowing faster than I walk). Now, it is possible to pick a θ such that $\alpha = 90°$. Since that α gives us a right triangle, we immediately have

$$\frac{v}{w} = \cos(\theta),$$

and so the turning angle from walking with the wind to walking so that the wind blows at right angles to my walking path is

$$\theta = \cos^{-1}\left(\frac{v}{w}\right).$$

For example, if I walk at 2 mph in a 5 mph wind, then my walking path should be at an angle of

$$\cos^{-1}\left(\frac{2}{5}\right) = 66.4°$$

to the wind's direction.

This problem is a serious illustration of how vectors can play an important role in mathematical physics, but to end this chapter on a lighter note, here's a "vector joke" for your amusement. What do you get when you cross a mosquito with a mountain climber? A biologist would surely say, "nothing, because that's impossible to do," and, strangely enough, a pure mathematician would not only agree but would claim to be able to *prove* the impossibility. Here's how.

In vector mathematics there are two different ways to multiply two vectors together: the *dot product* (which produces a scalar result), and the *cross product* (which produces another vector). Both types of product occur in physics, but each starts with two vectors. Notice, however, that while a mosquito is indeed a disease *vector*, a mountain climber is a scalar (groan). And you simply cannot *cross* a vector with a scalar. (This has to be an 11 on a scale of 10 in awful puns.)

Notes

1. A less self-centered interpretation can be found in R. L. Armstrong, "Relative Velocities and the Runner," *American Journal of Physics*, September 1978, pp. 950–951.

2. Passengers in a hot-air balloon basket that travels with the wind, *at the speed of the wind*, have $v = w$, and so $v' = 0$. That is, the people in the basket (traveling at what an observer on the ground might measure as a strong wind) *feel no wind at all*.

8. An Illuminating Problem

I bet Einstein turned himself all sorts of colors before he
invited the lightbulb.
— *Homer Simpson, once again proving he is an idiot*

Here you'll see how just simple algebra, combined with the physics of electric resistor circuits (Ohm's law and Kirchhoff's laws[1]), allows us to answer questions like the following. In Figures 8.1 and 8.2 we have two circuits, each made from ideal batteries,[2] incandescent lightbulbs, and a switch. In both circuits the batteries have identical voltages, and the bulbs are identical (in particular, their filaments have equal resistances). For both circuits we are to state how the brightness of each bulb changes from when the switch is open (as shown in the figures) to when the switch is closed. In addition, for the circuit in Figure 8.1 we are to answer the question again after the rightmost battery has been reversed.

For the circuit in Figure 8.1 with the switch open (as shown), clearly, the current in both bulbs is the same, $\frac{V}{R}$, where R is the resistance of each filament, and so each bulb has the same brightness. Once the switch is closed, we have the circuit shown in Figure 8.3, where each bulb has been replaced by its equivalent resistance R.

With reference to the ground node, the voltage at the top of the battery stack is $2V$, and the voltage at the node common to both bulbs is some value that we'll call E. Then, we can write the following equations:

$$I_1 = \frac{E}{R},$$

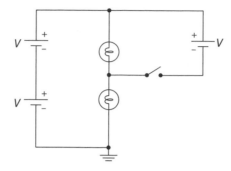

Figure 8.1. After the switch is closed, how does the brightness level of each bulb change? How does reversing the rightmost battery change your answer?

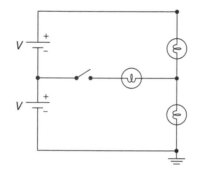

Figure 8.2. After the switch is closed, how does the brightness level of each bulb change?

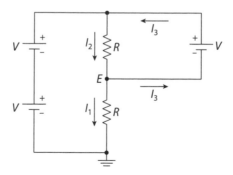

Figure 8.3. The circuit of Figure 8.1 with the switch closed

and

$$I_2 = \frac{2V - E}{R}.$$

Also, because of the rightmost battery,

$$E + V = 2V,$$

and so $E = V$. Thus,

$$I_1 = \frac{V}{R},$$

and

$$I_2 = \frac{2V - E}{R} = \frac{2V - V}{R} = \frac{V}{R}.$$

Notice that I_1 and I_2 are just what they were when the switch was open. So, there is *no change* in the brightness of either bulb. (Notice, too, that the current in the rightmost battery, I_3, is zero, because $I_2 = I_1 + I_3$ at the E-node.)

Again for the circuit of Figure 8.1, but now with the rightmost battery reversed, once the switch is closed we have the circuit of Figure 8.4. The equations for this circuit are

$$I_1 = \frac{E}{R}$$

and

$$I_2 = \frac{2V - E}{R},$$

where now, starting at the top of the battery stack on the left and moving through the right-most battery,

$$2V + V = E = 3V.$$

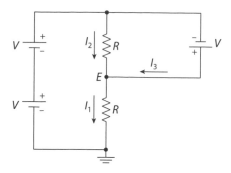

Figure 8.4. The circuit of Figure 8.1 with the switch closed and the rightmost battery reversed

So,

$$I_1 = \frac{3V}{R},$$

and

$$I_2 = \frac{2V - 3V}{R} = -\frac{V}{R}.$$

Thus, I_1 has tripled, and I_2 has reversed direction (but is unchanged in magnitude). This means that there is *no change* in the brightness of the upper bulb, but there is an *increase* in the brightness of the lower bulb.

Turning our attention now to the circuit of Figure 8.2, with the switch open the middle bulb is, of course, not lit at all, since current through it is zero, while the upper and lower bulbs have the same brightness, as the current through each is the same, $\frac{2V}{2R} = \frac{V}{R}$. Once the switch is closed we have the circuit of Figure 8.5, and its equations are

$$I_1 = \frac{V - E}{R}, \quad I_2 = \frac{2V - E}{R}, \quad I_3 = \frac{E}{R}.$$

Since

$$I_1 + I_2 = I_3,$$

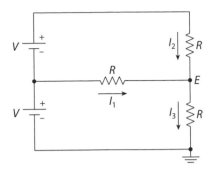

Figure 8.5. The circuit of Figure 8.2 with the switch closed

then

$$\frac{V - E}{R} + \frac{2V - E}{R} = \frac{E}{R},$$

or

$$V - E + 2V - E = E,$$

and so $E = V$. Thus,

$$I_1 = 0, \; I_2 = \frac{V}{R} = I_3,$$

and so there is *no change* in the brightness of any of the bulbs.

Some readers may scoff at the use of equations on these circuits, feeling that unleashing the power of math is overkill when one can just "look at the circuits and see'" the answers. I don't for an instant doubt that some readers can, in fact, do just that. Alas, however, even with a PhD in electrical engineering it doesn't take much to make *me* doubt my intuitive feelings about most circuits, and I do make that admission with regret. Even when I'm *sure* I know what's going to happen in a circuit, doing a formal analysis still makes me feel a lot better. So, let me challenge skeptical readers (who may still be snickering) with a variation on the circuit of Figure 8.2.

Suppose the upper bulb is replaced with one that has twice the filament resistance of the other two bulbs. What happens now when the switch is closed? Write down your answer, *right now*, before reading on.

With the switch open the middle bulb is of course unlit (as before), and the upper and lower bulbs have the same current in them (as before), equal now to $\frac{2V}{3R} = 0.67\frac{V}{R}$. Since the upper and lower bulbs have different filament resistances they won't be at the same brightness, but both bulbs *will* be lit.

With the switch closed the circuit equations now appear only subtly different from what they were before, but that difference has big repercussions:

$$I_1 = \frac{V - E}{R}, \quad I_2 = \frac{2V - E}{2R}, \quad I_3 = \frac{E}{R}.$$

Since

$$I_1 + I_2 = I_3,$$

then

$$\frac{V - E}{R} + \frac{2V - E}{2R} = \frac{E}{R}$$

or

$$2V - 2E + 2V - E = 2E$$

or

$$4V = 5E,$$

and so $E = \frac{4}{5}V$. Thus,

$$I_1 = \frac{V - \frac{4}{5}V}{R} = \frac{1}{5}\frac{V}{R} = 0.2\frac{V}{R}(\text{was } 0),$$

$$I_2 = \frac{2V - \frac{4}{5}V}{2R} = \frac{3}{5}\frac{V}{R} = 0.6\frac{V}{R}\left(\text{was } 0.67\frac{V}{R}\right),$$

$$I_3 = \frac{E}{R} = \frac{4}{5}\frac{V}{R} = 0.8\frac{V}{R}\left(\text{was } 0.67\frac{V}{R}\right).$$

So, closing the switch causes the middle bulb to go from unlit to glowing (though less intensely than was the identical lower bulb when the switch was open), the upper bulb to dim slightly (but it will still be brighter than is the now-glowing middle bulb), and the lower bulb will be a bit brighter than it was when the switch was open.

Now, *be honest*—is all that in what you wrote as your answer *before* doing the math?

Notes

1. Named after the German physicist Gustav Kirchhoff (1824–1887), the two laws are (a) the sum of the voltage drops around any closed loop is zero (this is an expression of the conservation of energy), and (b) the sum of all the currents into any node is zero (this is an expression of the conservation of electric charge). Ohm's law—named after the German physicist Georg Ohm (1789–1854)—is the well-known "voltage drop across a resistor is the product of the current in the resistor and the resistance."

2. An *ideal* battery has *zero* internal resistance. A *real* battery always has some positive internal resistance, typically quite small (a small fraction of an ohm) when new, that grows ever larger as the battery ages.

9. How to Measure Depth with a Stopwatch

In another moment down went Alice [after the white rabbit, into its hole], never once considering how in the world she was to get out again.
— *Lewis Carroll*, Alice's Adventures in Wonderland

To get you in the right mind-set for the main discussion, consider first this little puzzler: A stone falls the last half of the height of a wall in half a second. Ignoring air drag, how high is the wall?

Let t_1 be the time required for the first half (in distance) of the fall. Thus, if x is the height of the wall, then

$$\frac{1}{2}x = \frac{1}{2}gt_1^2,$$

and so

$$t_1 = \sqrt{\frac{x}{g}}.$$

If t_2 is the time required for the *entire* fall, then

$$x = \frac{1}{2}gt_2^2,$$

and thus

$$t_2 = \sqrt{\frac{2x}{g}}.$$

So,

$$t_2 - t_1 = \sqrt{\frac{2x}{g}} - \sqrt{\frac{x}{g}} = \frac{1}{2},$$

or

$$\sqrt{x}\left(\sqrt{\frac{2}{g}} - \sqrt{\frac{1}{g}}\right) = \frac{1}{2} = \sqrt{x}\,\frac{\sqrt{2} - \sqrt{1}}{\sqrt{g}} = \sqrt{x}\,\frac{\sqrt{2} - 1}{\sqrt{g}},$$

and so

$$\frac{1}{4} = x\left(\frac{\sqrt{2} - 1}{\sqrt{g}}\right)^2 = x\,\frac{\left(\sqrt{2} - 1\right)^2}{g}.$$

Therefore,

$$x = \frac{g}{4\left(\sqrt{2} - 1\right)^2} = \frac{32.2}{4\left(\sqrt{2} - 1\right)^2}\ \text{feet} = 46.9\ \text{feet}.$$

Notice—no quadratic equation is involved. We won't be able to get away quite so easily in the main question of this chapter.

Imagine that you are standing at the edge of a deep, vertical hole in the ground. It is *so* deep that the hole descends into blackness, with no bottom in sight. Deciding that you just have to know *how* deep it is, you think to use some simple physics to satisfy your curiosity. All you need is the small steel ball (the size of a marble) and the stopwatch you just happen to have in your pocket. So, here's what you do.

At the same instant you drop the steel ball into the hole you start the stopwatch. When you hear either a splash (or a clunk if the hole is dry) you stop the watch. Knowing the speed of sound is 1,115 feet/second, and ignoring air drag on the ball as it falls, determine how deep is the hole if the stopwatch reads 3 seconds. How deep would the hole be if the watch read 6 seconds? Explain why the calculated depth for a 6-second drop is not twice that for a 3-second drop.

Let t_1 be the time required for the ball to hit bottom, and t_2 be the time required for the sound of the ball arriving at the bottom to travel back up the hole to your ears. Thus, the *total* time (what the stopwatch reads) is T, where

$$T = t_1 + t_2.$$

If D is the depth of the hole, and if s is the speed of sound, then we know that (remember, we are ignoring the effect of air drag)

$$D = \frac{1}{2}gt_1^2$$

and that

$$t_2 = \frac{D}{s}.$$

So,

$$t_1 = \sqrt{\frac{2D}{g}},$$

and therefore

$$T = \sqrt{\frac{2D}{g}} + \frac{D}{s},$$

or

$$sT - D = s\sqrt{\frac{2D}{g}}.$$

Squaring both sides and collecting terms, we arrive at the following quadratic equation for D:

$$D^2 - \left(\frac{2s^2}{g} + 2sT\right)D + s^2T^2 = 0.$$

The well-known solution for a quadratic equation then gives us our answer:

$$D = \frac{s^2 + sTg \pm s^2\sqrt{1 + 2\frac{Tg}{s}}}{g}.$$

Indeed, we have an embarrassment of riches, with *two* solutions for D because of the $\pm$ sign. Of course *both* of these solutions can't be right, so which one do we keep and which one do we reject? We can decide that question by seeing what our two candidate solutions do for the extreme case of $T = 0$. *Physically* we know that means $D = 0$. If we use the plus sign we get $D = \frac{2s^2}{g}$, which is clearly wrong. If we use the minus sign, however, we get $D = 0$. So, the depth of the hole is

$$D = \frac{s^2 + sTg - s^2\sqrt{1 + 2\frac{Tg}{s}}}{g} = \frac{s^2\left[1 - \sqrt{1 + 2\frac{Tg}{s}}\right] + sTg}{g},$$

or, finally,

$$D = sT + \frac{s^2}{g}\left[1 - \sqrt{1 + 2\frac{Tg}{s}}\right].$$

For a 3-second drop the depth of the hole is

$$D = (1,115)3 + \frac{1,115^2}{32.2}\left[1 - \sqrt{1 + 2\frac{3(32.2)}{1,115}}\right] \text{ feet}$$

$$= [3,345 + 38,609\,(-0.08318)] \text{ feet}$$

$$= [3,345 - 3,211] \text{ feet} = 134 \text{ feet}.$$

For a 6-second drop the depth of the hole is

$$D = (1,115)6 + \frac{1,115^2}{32.2}\left[1 - \sqrt{1 + 2\frac{6(32.2)}{1,115}}\right] \text{ feet}$$

$$= [6,690 + 38,609\,(-0.1604)] \text{ feet}$$

$$= [6,690 - 6,193] \text{ feet} = 497 \text{ feet},$$

which is a *lot* more than simply double the depth of a 3-second drop. Here's why.

At the end of a 6-second drop the ball is moving at the speed $gt = 32.2\,(6)$ feet/second $= 193$ feet/second, which is much less than the speed of sound. So, most of the 6 seconds is used for the drop itself, with only a small part of the 6 seconds required for the sound of the ball's arrival at the bottom to travel back up the hole. That is, the ball drops at an ever-increasing speed for almost all of the 6 seconds and so travels for greater than double the distance traveled during a 3-second drop.

10. Doing the Preface Problems

Love makes the world go round.

—*popular saying that admittedly loses something when restated as*
"angular momentum makes the world spin"

In earlier chapters we encountered the concepts of moment of inertia and torque, and here I'll extend those discussions to allow us to answer two of the questions I left you with at the end of the preface (along with the question on the Moon's recession from the Earth that I mentioned at the end of Example 6 in Chapter 1). To start, let me remind you of a few things.

If the mass m is moving with speed v, then it has a kinetic energy of linear (straight-line) motion given by

$$E_{\text{linear}} = \frac{1}{2}mv^2.$$

In Chapter 6 we found that even if a mass isn't in linear motion but is instead spinning in place with an angular rotation speed of Ω radians/second, it has a *rotational* kinetic energy given by

$$E_{\text{rotation}} = \frac{1}{2}I\Omega^2,$$

where I is the *moment of inertia* of the mass around the spin axis. These two boxed energy expressions suggest the following analogy: I is "like" m, and Ω is "like" v. So, if we extend the analogy to momentum, then, since the linear momentum is mv, we might reasonably say that the *angular momentum* is $I\Omega$.

We can use the two boxed energy expressions to answer the preface questions about the two cylinders (one hollow and the other solid) rolling down an inclined plane (take another look at Figure P1). At time $t = 0$ both cylinders, of equal mass m and radius R, are at rest distance L up the incline (which is at angle θ to the horizontal). Thus, both cylinders initially have *zero* kinetic energy and a potential energy of $mgL \sin(\theta)$. When each cylinder has rolled distance x down the incline it will have traded some of that potential energy—$mgx \sin(\theta)$— for kinetic energies of linear *and* rotational motion. That is, if I, $\Omega(x)$, and $v(x)$ are the moment of inertia, the spin rate, and the speed of linear motion down the incline of a cylinder, respectively, when the cylinder is distance x down the incline, then

$$\frac{1}{2}mv^2 + \frac{1}{2}I\Omega^2 = mgx \sin(\theta).$$

If $T(x)$ is the time for one revolution of the cylinder at distance x, then

$$\Omega(x) = \frac{2\pi}{T(x)},$$

and so

$$T(x) = \frac{2\pi}{\Omega(x)}.$$

Since a cylinder rolls down the incline a distance of $2\pi R$ for each revolution, we have

$$v(x) = \frac{2\pi R}{T(x)} = \frac{2\pi R}{\frac{2\pi}{\Omega(x)}} = \Omega(x)R,$$

and so

$$\Omega(x) = \frac{v(x)}{R}.$$

Substituting this expression for Ω into the last boxed equation, we have

$$\boxed{\frac{1}{2}mv^2 + \frac{1}{2}I\frac{v^2}{R^2} = mgx\,\sin(\theta).}$$

This last boxed expression is true for both cylinders, in general, but of course, I is different for hollow and solid cylinders. Let's now treat each cylinder in turn, starting with the solid cylinder.

As we showed in Chapter 6,

$$I_{\text{solid}} = \frac{1}{2}mR^2,$$

and so

$$\frac{1}{2}mv^2 + \frac{1}{4}mv^2 = mgx\,\sin(\theta) = \frac{3}{4}mv^2,$$

or

$$v^2 = \frac{4gx\,\sin(\theta)}{3}.$$

Since

$$v = \frac{dx}{dt},$$

we therefore have

$$\frac{dx}{dt} = \sqrt{\frac{4g\,\sin(\theta)}{3}}\sqrt{x},$$

or, integrating from $0 \le x \le L$ (and so $0 \le t \le t_{\text{solid}}$, where t_{solid} is the time it takes the solid cylinder to reach the bottom of the incline),

$$\int_0^L \frac{dx}{\sqrt{x}} = \int_0^{t_{\text{solid}}} \sqrt{\frac{4g \sin(\theta)}{3}}\, dt = 2t_{\text{solid}}\sqrt{\frac{g \sin(\theta)}{3}} = 2\left(\sqrt{x}\right)\big|_0^L = 2\sqrt{L}.$$

Thus,

$$t_{\text{solid}} = \sqrt{\frac{3L}{g \sin(\theta)}}.$$

Now, let's repeat these calculations for a hollow cylinder. We also know from Chapter 6 that

$$I_{\text{hollow}} = m R^2,$$

and so

$$\frac{1}{2}m v^2 + \frac{1}{2}m v^2 = mgx \sin(\theta) = m v^2.$$

Thus,

$$v^2 = gx \sin(\theta),$$

and so

$$\frac{dx}{dt} = \sqrt{g \sin(\theta)}\sqrt{x}.$$

Thus, if t_{hollow} is the time it takes the hollow cylinder to reach the bottom of the incline, we have

$$\int_0^L \frac{dx}{\sqrt{x}} = \int_0^{t_{\text{hollow}}} \sqrt{g \sin(\theta)}\, dt = t_{\text{hollow}}\sqrt{g \sin(\theta)} = 2\sqrt{L},$$

or

$$t_{\text{hollow}} = \sqrt{\frac{4L}{g \sin(\theta)}}.$$

So, the *solid* cylinder wins the race down the incline, as $t_{\text{hollow}} > t_{\text{solid}}$. Our calculations even tell us by how much the solid cylinder wins, as we have

$$\frac{t_{\text{hollow}}}{t_{\text{solid}}} = \sqrt{\frac{4L}{g \sin(\theta)}} \sqrt{\frac{g \sin(\theta)}{3L}} = \sqrt{\frac{4}{3}} = \frac{2}{\sqrt{3}} = 1.1547.$$

That is, the hollow cylinder takes a bit more than 15% longer to roll to the bottom of the incline than does the solid cylinder.

Of what practical value is this knowledge, you ask? Well, suppose you are in a county fair contest in which a blue ribbon goes to the person who rolls down an incline the fastest—*using a barrel*. (I've seen even odder things than *that* at county fairs!) Our result shows that if you stuff yourself *inside* the barrel you'll do better than if you wrap yourself around the *outside* of the barrel. Somehow, I think you'd go with the first option anyway, but now you know it's the right choice according to theoretical physics as well as to common sense!

Let's now turn our attention to another preface challenge question, that of the falling chimney (see Figures P2 and P3 again). To start, I'll first establish a most useful relationship linking torque, moment of inertia, and angular acceleration. We begin by imagining a point mass m with a force F acting on it to produce an acceleration of a. Thus, $F = ma$, or

$$a = \frac{F}{m}.$$

If this point mass is moving at angular speed Ω radians/second on a circular path of radius r, then its tangential speed is $v = r\Omega$.

If Ω changes by $\Delta\Omega$, then v changes by Δv, and so

$$v + \Delta v = r(\Omega + \Delta\Omega),$$

or

$$\Delta v = r\Delta\Omega.$$

If Δv and $\Delta\Omega$ occur in Δt time, then

$$\boxed{\frac{\Delta v}{\Delta t} = r\frac{\Delta\Omega}{\Delta t},}$$

and so, in the limit $\Delta t \to 0$, the point mass experiences *angular* acceleration

$$\lim_{\Delta t \to 0} \frac{\Delta\Omega}{\Delta t} = \alpha$$

and *tangential* acceleration

$$\lim_{\Delta t \to 0} \frac{\Delta v}{\Delta t} = a.$$

Thus, from the last boxed expression we have

$$a = r\alpha,$$

or

$$\alpha = \frac{a}{r} = \frac{\frac{F}{m}}{r} = \frac{F}{mr} = \left(\frac{r}{r}\right)\frac{F}{mr} = \frac{rF}{mr^2}.$$

Recall from Chapter 6 that we called the product rF the *torque*, and mr^2 the moment of inertia of a point mass m that is distance r from a rotation center. That is,

$$\text{angular acceleration} = \frac{\text{torque}}{\text{moment of inertia}},$$

or

> torque = (moment of inertia)(angular acceleration).

Now, back to Figure P2. As the chimney falls and before any buckling occurs (*if* it occurs), we have two equal point masses traveling on circular paths. The point mass at b is on a circular path of radius L, and the point mass at c is on a circular path of radius $2L$. Since the two masses are equal, the components of their weights perpendicular to the length of the chimney are also equal (let's call those components F_b and F_c). The torques that these components exert (about the pivot point at the bottom of the chimney, the point the chimney is rotating around) are given by $T_b = F_b L$ and $T_c = F_c 2L$, or, since $F_b = F_c$, we have

$$T_c = 2T_b.$$

The moment of inertia of the point mass at b, about the pivot point, is $I_b = mL^2$, while the moment of inertia of the point mass at c, about the pivot point, is $I_c = m4L^2$. That is,

$$I_c = 4I_b.$$

If we call α_b and α_c the angular accelerations of the masses at b and c, respectively, then when we substitute our results for T and I into the last boxed expression we get

$$T_b = I_b \alpha_b,$$

and

$$T_c = I_c \alpha_c.$$

That is,

$$\alpha_b = \frac{T_b}{I_b},$$

and

$$\alpha_c = \frac{T_c}{I_c}.$$

Thus,

$$\frac{\alpha_b}{\alpha_c} = \frac{\frac{T_b}{I_b}}{\frac{T_c}{I_c}} = \left(\frac{T_b}{T_c}\right)\left(\frac{I_c}{I_b}\right) = \left(\frac{1}{2}\right)(4) = 2.$$

That is, the point mass at b has twice the angular acceleration as does the point mass at c, and so it acquires angular speed more rapidly than does the point mass at c. Thus, our simple chimney model *does* buckle and does so as shown in Figure P3(a). Photographs of falling chimneys show that this is, indeed, the way real chimneys buckle as they fall.[1]

Finally, we end this chapter with a most impressive illustration of simple physics at work. Recall that in Example 6 of Chapter 1 I told you that laser pulse measurements, using the corner cube reflectors placed on the Moon by the *Apollo 11* astronauts, show that the Moon's distance from Earth is increasing at the rate of about an inch and a half per year. I'll now show you how to *calculate* that recession rate using the conservation of angular momentum, one of the fundamental laws of physics.

We start by imagining the Earth/Moon system as all alone in the Universe, with only the distant stars as a backdrop. The Moon revolves around a rotating Earth, an Earth that is otherwise stationary in space with respect to those distant stars. This is, of course, not the real situation, but it greatly simplifies the analysis while retaining enough of reality to keep the physics honest. Since the Earth is rotating, it has spin angular momentum $I\Omega$, as stated at the beginning of the chapter, and since the Moon is revolving about the Earth, it has orbital angular momentum (which we'll work out in just a moment).

We established in Chapter 6 that the Earth's rotation rate is slowing, because of tidal friction. This means the Earth's spin angular momentum is decreasing. Since angular momentum is conserved, somewhere else in our Earth/Moon system angular momentum must be increasing. Where's that "somewhere"? The only other place is the Moon: the ocean tides cause a transfer of spin angular momentum from the Earth to the Moon, in particular to the Moon's *orbital* angular momentum. One might imagine that the Moon's spin angular momentum could increase, too, but that is not observed to occur. I'll assume that the Moon's orbital angular momentum is the only beneficiary of Earth's loss of spin angular momentum, and we'll see where that takes us.[2]

Now, what is the Moon's orbital angular momentum? We imagine that the Moon, which we'll take as a point mass m, orbits the Earth in a circular path of radius r with a speed v. If the Moon's angular speed is Ω radians/second, then

$$v = \omega r,$$

and so

$$\omega = \frac{v}{r}.$$

Also, as it orbits the Earth the Moon's moment of inertia is

$$I = mr^2,$$

and since the Moon's orbital angular momentum is $I\omega$, we can write the orbital angular momentum as

$$L_m = I\omega = mr^2 \left(\frac{v}{r}\right) = mrv.$$

The units of angular momentum are $\frac{\text{kilograms·meters-squared}}{\text{seconds}}$. Notice carefully (if you haven't already) that *linear* momentum (mv) and *angular* momentum (mrv) have different units. This result shouldn't be too shocking, however, as we've already seen a similar situation with the different units for *orbital* speed (v) and *angular* speed (ω).

We are now ready to calculate the Moon's recession rate. Let M be the Earth's mass. As the gravitational force on the Moon by the Earth is

$$F = \frac{GMm}{r^2},$$

then if we set the gravitational acceleration of the Moon equal to its centripetal acceleration, we have

$$\frac{F}{m} = \frac{GM}{r^2} = \frac{v^2}{r},$$

and so

$$v = \sqrt{\frac{GM}{r}}.$$

This means that the orbital angular momentum of the Moon is

$$L_m = mr\sqrt{\frac{GM}{r}} = m\sqrt{GM}\sqrt{r}.$$

Differentiating with respect to r, we obtain

$$\frac{dL_m}{dr} = m\sqrt{GM}\frac{1}{2}\frac{1}{\sqrt{r}},$$

or if we approximate differentials with delta changes,

$$\Delta r \approx \frac{2}{m}\sqrt{\frac{r}{GM}}\Delta L_m.$$

That is, a positive change ΔL_m in the Moon's orbital angular momentum leads to a positive change Δr in its orbital radius.

The central assumption in this analysis is that the ΔL_m change is equal in magnitude to the change ΔL_e in the Earth's spin angular momentum. In Chapter 6 we found that the Earth's moment of inertia is $0.3444MR^2$, where R is the Earth's radius. Thus, the spin angular momentum of the Earth is

$$L_e = 0.3444MR^2\Omega,$$

where Ω is the Earth's rotation rate. Now, if T is the length of a day (86,400 seconds), then

$$\Omega = \frac{2\pi}{T} \text{ radians/second,}$$

and since

$$\Delta L_e = 0.3444 M R^2 \Delta\Omega,$$

and since

$$\frac{d\Omega}{dT} = -\frac{2\pi}{T^2}$$

says (for small changes) that

$$\Delta\Omega = -\frac{2\pi}{T^2}\Delta T,$$

then

$$\Delta L_e = -0.3444 M R^2 \frac{2\pi}{T^2}\Delta T.$$

In these last two expressions ΔT is the change in the length of a day that is associated with the change in Earth's rotation rate in time interval T. Recall that in Chapter 6 we found T changes 0.002 seconds in 100 years, and so the *daily* change is

$$\Delta T = \frac{2\times 10^{-3}\ \text{seconds}}{(100\,\text{years})\left(365\frac{\text{days}}{\text{year}}\right)} = \frac{2\times 10^{-5}}{365}\frac{\text{seconds}}{\text{day}}.$$

Thus, the *daily* change in Earth's spin angular momentum is

$$\Delta L_e = -\frac{0.6888 M R^2 \pi}{(86,400\ \text{seconds})^2}\left(\frac{2\times 10^{-5}}{365}\frac{\text{seconds}}{\text{day}}\right).$$

We then multiply by 365 days to get the *yearly* change in ΔL_e:

$$\boxed{\Delta L_e = -\frac{0.6888 M R^2 \pi}{86,400^2}2\times 10^{-5}\frac{1}{\text{seconds}}.}$$

Using $\Delta L_m = |\Delta L_e|$ in the boxed expression for Δr, we get the *yearly* change in the Moon's orbital radius:

$$\Delta r = \frac{2}{m}\sqrt{\frac{r}{GM}\frac{0.6888MR^2\pi}{86,400^2}}\, 2 \times 10^{-5}\frac{1}{\text{seconds}},$$

or

$$\boxed{\Delta r = \frac{4\pi(0.6888)R^2}{86,400^2 m}\sqrt{\frac{Mr}{G}} \times 10^{-5}\frac{1}{\text{seconds}}.}$$

Evaluating this expression for Δr should give us a result in units of meters. To check that this is so, let's explicitly insert all the units for all the entries in the boxed expression:

m = mass of Moon = 7.35×10^{22} kilograms,
M = mass of Earth = 5.98×10^{24} kilograms,
r = radius of Moon's orbit = 239,000 miles = 3.84×10^8 meters,
G = gravitational constant = $6.67 \times 10^{-11} \frac{\text{meters cubed}}{\text{kilograms·seconds-squared}}$,
R = Earth's radius = 6.37×10^6 meters
and so

$$\Delta r = \frac{4\pi(0.6888)(6.37 \times 10^6 \text{ meters})^2}{(8.64 \times 10^4)^2(7.35 \times 10^{22} \text{ kilograms})}$$

$$\sqrt{\frac{(5.98 \times 10^{24} \text{ kilograms})(3.84 \times 10^8 \text{ meters})}{6.67 \times 10^{-11}\frac{\text{meters squared}}{\text{kilograms·seconds-squared}}}} \times 10^{-5}\frac{1}{\text{seconds}}$$

$$= 0.64 \times 10^{-18}\frac{\text{meters squared}}{\text{kilograms}}$$

$$\times\sqrt{3.44 \times 10^{43}\frac{\text{kilograms-squared · seconds-squared}}{\text{meters squared}}}$$

$$\times 10^{-5}\frac{1}{\text{seconds}}$$

$$= 0.64 \times 10^{-23} \frac{\text{meters squared}}{\text{kilograms} \cdot \text{seconds}}$$

$$\times \sqrt{34.4 \times 10^{42} \frac{\text{kilograms-squared} \cdot \text{seconds-squared}}{\text{meters squared}}}$$

$$= 3.75 \times 10^{-23} \text{ meters} \times 10^{21}$$

$$= 3.75 \times 10^{-2} \text{ meters} = 3.75 \text{ centimeters}.$$

Remember, this is the *yearly* change, and since there are 2.54 centimeters in an inch, we get a recession rate of 1.48 inches/year, a theoretically calculated value in *outstanding* agreement with the laser/corner cube experimental measurements.

Notes

1. See Francis B. Bundy, "Stresses in Freely Falling Chimneys and Columns," *Journal of Applied Physics*, February 1940, pp. 112–123 (in particular, p. 121).

2. The relationship between the Earth and the Moon is enormously complicated, one not in any sense of the word describable as being "simple." An old but still extremely useful introduction to the subject is Gordon J. F. MacDonald, "Earth and Moon: Past and Future," *Science*, August 28, 1964, pp. 881–890. MacDonald observes that the recession rate has been almost constant over the past one billion years, and so a billion years ago the Moon was about 1.5 billion inches (23,600 miles) closer to the Earth than it is today.

11. The Physics of Stacking Books

Every miser knows that a stack of pennies can be "leaned"
slightly off vertical without falling. How far can the top
penny be from its position in a vertical stack?"
— *Paul B. Johnson*[1]

The epigraph describes a situation that never fails to astonish all
who first see it. Johnson answered his penny question by deriving
a mathematical equation and solving it with some subtle arguments.
Here I'll do it using just some simple physics, in which the concept of
the *center of mass* of a spatially extended object will play an important
role. The center of mass is the point at which we can imagine the entire
mass of the object is concentrated as a *point* mass. Often, the center
of mass is obvious by inspection because of symmetry. For example,
the center of mass of a uniformly dense solid sphere is the geometric
center of the sphere. Similarly, the center of mass of a circular hoop
with uniform density is the center of the hoop (but notice there is,
in this case, no mass actually at the center of mass). If the extended
object is at all complicated, and symmetry arguments fail, then the
center of mass has to be calculated. In the simplest case, suppose we
have N point masses, m_i, $1 \leq i \leq N$, located at (x_i, y_i, z_i). Then, the
x-coordinate of the center of mass is given by

$$X_C = \frac{\sum_{i=1}^{N} m_i x_i}{\sum_{i=1}^{N} m_i},$$

and similar expressions hold for Y_C and Z_C.

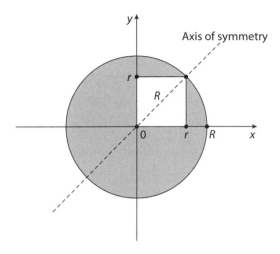

Figure 11.1. A circular disk with a square removed

Sometimes, when symmetry might *appear* to be absent, it really isn't. An example of this—a favorite of freshman physics teachers who need an exam question on short notice—is shown in Figure 11.1. There you see a circular disk of uniform thickness and density, with the largest possible square cut out of the upper-right quadrant. When the disk was still intact, symmetry told us that its center of mass was at the origin. With the square removed, however, that's no longer the case— and that's the question: where *is* the center of mass for the *cut* disk? Let's call the answer to that question (X, Y). Now, even with the cut there is still enough symmetry left in the disk to argue that $Y = X$ (that is, as the "axis of symmetry" shown in Figure 11.1 indicates, there is nothing to distinguish the x- and the y-directions). That observation helps a bit, but we are still left with the question, what is X?

The center of mass of the square cut from the disk is, by symmetry, in the middle of the square. From simple geometry (remember the Pythagorean theorem), if the radius of the disk is R, then the edge length of the square is $r = \frac{R}{\sqrt{2}}$, and so the center of the square is at $(\frac{R}{2\sqrt{2}}, \frac{R}{2\sqrt{2}})$. Now, here's the crucial observation: if we put the square back into the cut disk, we get the *intact* disk back. Who could argue with that? So, if m_1 is the mass of the cut disk and if m_2 is the mass of the square, then the formula for the center of mass resulting from combining two individual masses says

$$0 = \frac{m_1 X + m_2 \frac{R}{2\sqrt{2}}}{m_1 + m_2}.$$

The zero on the left is because, as argued by symmetry, that's the x-coordinate of the center of mass of the once-again intact disk. So, just like that, we have

$$X = -\frac{m_2}{m_1}\left(\frac{R}{2\sqrt{2}}\right).$$

Or, since the disk and the square are of uniform thickness and density, the masses of these two objects are directly proportional to their surface areas (A_1 and A_2, respectively), we can write

$$X = -\frac{A_2}{A_1}\left(\frac{R}{2\sqrt{2}}\right).$$

From geometry we have

$$A_1 = \pi R^2 - A_2,$$

and

$$A_2 = \frac{R^2}{2}.$$

So,

$$X = -\frac{\frac{R^2}{2}}{\pi R^2 - \frac{R^2}{2}}\left(\frac{R}{2\sqrt{2}}\right) = -\frac{R}{(2\pi - 1)2\sqrt{2}} = -0.06692R(=Y).$$

Isn't that slick? Okay, now that you see how the center of mass formula works, off we go to the real topics of this chapter.

Instead of Johnson's pennies (you'll see why in just a bit), imagine a book of length 1 and mass 1 lying flat on a tabletop with the book's rightmost edge right at the edge of the table, as shown in Figure 11.2. The left edge of the book is at $x = 0$, and so the right edge of the book (and the edge of the table) is at $x = 1$. The center of mass of the book is at $x = \frac{1}{2}$, and so we can slide the book forward distance $\frac{1}{2}$ before the book will fall off the table. The book projects out beyond the tabletop by $\frac{1}{2}$, and that projection is called the *overhang*, denoted by S. So, for one book, we have $S(1) = \frac{1}{2} = \frac{1}{2}(1)$.

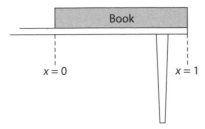

Figure 11.2. A book lying flat on a tabletop

Now, imagine two such books neatly stacked on the table. From our first analysis we know we can slide the top book forward distance $\frac{1}{2}$ before it falls off the bottom book. The center of mass of the top book is now at $x = 1$. The center of mass of the two books together is at

$$x = \frac{1\left(\frac{1}{2}\right) + 1(1)}{2} = \frac{3}{4},$$

and so we can slide the two-book combo forward distance $\frac{1}{4}$ toward the table edge before the combo falls off the table. Now, the projection of the top book beyond the table edge is

$$S(2) = \frac{1}{4} + \frac{1}{2} = \frac{1}{2}\left(1 + \frac{1}{2}\right).$$

Let's do this just one more time, with three identical books neatly stacked on the table. From our earlier results we know we can slide the top book forward by distance $\frac{1}{2}$ before it falls off the middle book, and then we can slide the upper two-book combo forward by distance $\frac{1}{4}$ before the combo falls off the bottom book. The center of mass of the upper two-book combo is now at $x = 1$. The center of mass of the three-book combo is at

$$x = \frac{1\left(\frac{1}{2}\right) + 2(1)}{3} = \frac{5}{6},$$

and so we can slide the three-book combo forward by distance $\frac{1}{6}$ toward the table edge before the combo falls off the table. Now, the projection of the top book beyond the table edge is

$$S(3) = \frac{1}{6} + \frac{1}{4} + \frac{1}{2} = \frac{1}{2}\left(1 + \frac{1}{2} + \frac{1}{3}\right).$$

By now you have probably begun to suspect that, in general, if we keep doing this, stacking ever more books, we'll find that

$$S(n) = \frac{1}{2}\sum_{k=1}^{n}\frac{1}{k}.$$

We can verify this suspicion by induction. That is, let's *suppose* that for $n-1$ books,

$$S(n-1) = \frac{1}{2}\sum_{k=1}^{n-1}\frac{1}{k},$$

and then we'll show that this implies

$$S(n) = \frac{1}{2}\sum_{k=1}^{n}\frac{1}{k}.$$

That would mean, since we've already shown by *direct calculation* that our supposed formula for $S(n)$ holds for $n = 3$, that it must hold for $n = 4$ (which means it holds for $n = 5$, and so on). We also know by direct calculation that our formula holds for $n = 1$ and $n = 2$, as well, of course.

So, before the final adjustment of the bottom book (and all the other books above it), the top $n-1$ books have their combined center of mass at $x = 1$ just before they will fall off the bottom book. The top book has a projection of $S(n-1)$ beyond the edge of the table. The center of mass of the n-book combo is at

$$x = \frac{1\left(\frac{1}{2}\right) + (n-1)(1)}{n} = \frac{1}{2n} + \frac{n-1}{n} = \frac{1+2(n-1)}{2n} = \frac{2n-1}{2n} = 1 - \frac{1}{2n}.$$

Thus, we can slide the n-book combo forward distance $\frac{1}{2n}$ toward the table edge before the n-book combo falls off the table. So,

$$S(n) = S(n-1) + \frac{1}{2n} = \frac{1}{2}\sum_{k=1}^{n-1}\frac{1}{k} + \frac{1}{2n} = \frac{1}{2}\sum_{k=1}^{n}\frac{1}{k},$$

just as we suspected, and our proof by induction is done.

Now, here's the "surprise." How big can $S(n)$ be? Answer: as big as you like! That's because $S(n)$ is the truncated form of the so-called harmonic series, which is well known to blow up as $n \to \infty$.[2] As the Russian-born physicist George Gamow (1904–1968) stated in one of his books when discussing this problem:[3] "By stacking an unlimited number of books ...we can make the top book protrude any desired distance beyond the edge of the table." His very next statement, though, was far off the mark: "Because of the rapidly decreasing contribution of each new book, however, we will need the entire Library of Congress to make the overhang equal to three or four book lengths!" That is not so.

It is quite easy to program a computer to evaluate $S(n)$ for given values of n; in fact, $S(n)$ first exceeds 3 when $n = 227$, and $S(n)$ first exceeds 4 when $n = 1{,}674$. Neither value of n is anywhere near the number of books in the Library of Congress. It's an entirely different story for larger values of $S(n)$, however: the overhang $S(n)$ first exceeds 50, for example, when n is something more than 1.5×10^{43}. Now *that* is many more books than are in the Library of Congress![4]

The appearance of Paul Johnson's note on the penny-stacking problem in the *American Journal of Physics* (note 1) prompted the following reply from a physicist at The Ohio State University who had solved the problem himself some years before: "To prove [the overhang] result 'physically,' a fellow graduate student and I stacked bound volumes of *The Physical Review* one evening, until an astonishingly large offset was obtained and left them to be discovered the next morning by a startled physics librarian."[5] Who says physicists are mostly shy, quiet nerds? In my book—and as Eisner's letter demonstrates—some of them are really crazy-wild guys!

Before leaving the general topic of center of mass, I'll end this chapter by showing you a somewhat more serious application than building stacks of offset pennies and books, namely, a dramatic

Figure 11.3. A domino chain reaction

illustration of the exponential (indeed, *explosive*) growth of energy in a chain reaction. To model neutrons successively splitting atomic nuclei, as occurs in atomic fission bombs, we'll use a falling domino to knock over a slightly larger domino, which will then knock over an even larger domino, and so on[6] (unlike the dominos in Figure 11.3, which are all the same size). The input energy required to knock over the initial domino can be quite small, while the energy released by the final falling domino can be *billions* of times larger (we'll prove this in just a bit). You can find videos of such domino chain reactions on YouTube, but they are strictly for fun viewing. Here I'll show you how to *calculate* the energies involved, using simple physics.

The communication in note 6 describes a chain of 13 ever-larger dominos, all made from acrylic plastic, with the smallest one (domino #1) having the dimensions

thickness $(w) = 1.19 \times 10^{-3}$ meters
width $(l) = 4.76 \times 10^{-3}$ meters
height $(h) = 9.53 \times 10^{-3}$ meters

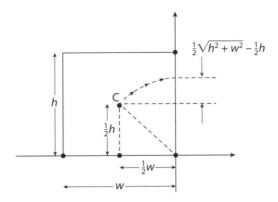

Figure 11.4. The geometry of an upright domino

and the largest one (domino #13) with the dimensions

thickness $(w) = 76.2 \times 10^{-3}$ meters
width $(l) = 305 \times 10^{-3}$ meters
height $(h) = 610 \times 10^{-3}$ meters

Starting with the smallest domino, each subsequent domino in the chain is slightly less than 1.5 times larger in each dimension than the previous one; it was stated in note 6 that the energy required to knock domino #1 over is 0.024×10^{-6} joules (see note 4 in Chapter 3 again), and the energy released by the fall of domino #13 is about 51 joules, an energy amplification factor of about 2 billion! The author of note 6 said: "It is easy to calculate [these energies]" but didn't show how to do it. So, let calculate them for ourselves.

Figure 11.4 shows a cross section of a domino, with its front face on the y-axis and its lower front edge at the origin (you are to imagine that the width, or l-dimension, is perpendicular to the page). The center of mass, C, of the domino is, by symmetry, located at the combined midpoints of each of the three dimensions. Imagine now that a force is applied to the left face of the domino. The domino will start to rotate clockwise round the lower front edge, and the center of mass will clearly rise until it is directly over the front edge. Any further rotation of the domino will place C beyond the front edge, and the domino will then topple over.

When C is directly above the front edge it will have elevated through the distance

$$\sqrt{\left(\frac{h}{2}\right)^2 + \left(\frac{w}{2}\right)^2} - \frac{h}{2} = \frac{h}{2}\sqrt{1 + \left(\frac{w}{h}\right)^2} - \frac{h}{2} = \frac{h}{2}\left[\sqrt{1 + \left(\frac{w}{h}\right)^2} - 1\right].$$

Thus, the potential energy of the domino increases by

$$\Delta E = mg\,\Delta y = mg\frac{h}{2}\left[\sqrt{1 + \left(\frac{w}{h}\right)^2} - 1\right],$$

where m is the mass of the domino. ΔE is the required input energy to topple the domino. The mass m is

$$m = \rho wlh,$$

where ρ is the density of acrylic plastic. A quick search on the Web gave the value of ρ as somewhere between 1.15 and 1.2 grams/cubic centimeter; I'll use an average of 1.18 grams/cubic centimeter $= 1.18 \times 10^3 \frac{\text{kilograms}}{\text{cubic meter}}$. So, for domino #1, the mass is

$$m = 1.19 \times 4.76 \times 9.53 \times 10^{-9} \text{ cubic meters} \times 1.18 \times 10^3 \frac{\text{kilograms}}{\text{cubic meter}}$$

$$= 63.7 \times 10^{-6} \text{ kilograms},$$

and therefore

$$\Delta E = \frac{1}{2}63.7 \times 10^{-6} \text{ kilograms} \times 9.8\frac{\text{meters}}{\text{seconds squared}}$$

$$\times 9.53 \times 10^{-3} \text{ meters} \left[\sqrt{1 + \left(\frac{1.19 \times 10^{-3}}{9.53 \times 10^{-3}}\right)^2} - 1\right]$$

$$= 2,975 \times 10^{-9}\frac{\text{kilograms} \cdot \text{meters-squared}}{\text{seconds squared}}(0.00777)$$

$$= 23 \times 10^{-9} \text{ joules}$$

$$= 0.023 \times 10^{-6} \text{ joules},$$

which is very close to the value declared by the author of note 6 (who suggested that this really quite small energy input could "be supplied by nudging [the domino] with a long wispy piece of cotton batton.")

Finally, to compute the energy released by the toppling of the largest domino, we start with its initial energy and then add the energy required to raise its center of mass to the point where it is over the domino's front edge. We then subtract the potential energy retained by the domino after it has fallen over. The result is the energy released by the domino. So, when domino #13 is upright its center of mass is at height 305×10^{-3} meters. When it's hit by domino #12 the center of mass of domino #13 rises to a height of

$$\frac{1}{2}\sqrt{(610)^2 + (76.2)^2} \times 10^{-3} \text{ meters} = 307.4 \times 10^{-3} \text{ meters}.$$

When domino #13 has toppled, the original w dimension is the new h dimension, and so the center of mass is at height 38.1×10^{-3} meters. The change (decrease) in the potential energy of the domino is therefore

$$mg\Delta y = \rho w l h g \Delta y$$

$$= 1.18 \times 10^3 \frac{\text{kilograms}}{\text{cubic meter}} \times 9.8 \frac{\text{meters}}{\text{seconds squared}}$$

$$\times 305 \times 76.2 \times 610 \times 10^{-9} \text{ cubic meters}$$

$$\times (307.4 - 38.1) \times 10^{-3} \text{ meters} = 44 \text{ joules}.$$

This result is "close" to 51 joules but still far enough off to warrant some concern. My guess is that the author of note 6 simply did a rough calculation and ignored the fact that the toppled center of mass was actually not at zero height. That is, he did the $mg\Delta y$ calculation but used 307.4×10^{-3} meters for Δy, which would result in a potential energy decrease of 50.4 joules.

The energy amplification factor achieved by the 13 falling dominos is, by the calculations here, the quite impressive value of

$$\frac{44}{0.023 \times 10^{-6}} = 1.9 \times 10^9 = 1.9 \, billion!$$

Notes

1. These are the opening words in Johnson's cleverly titled note that simultaneously alludes to Italian money and that country's famous tower in Pisa: "Leaning Tower of Lire," *American Journal of Physics*, April 1955, p. 240.

2. Here's a simple demonstration of that:

$$\lim_{n \to \infty} S(n) = 1 + \frac{1}{2} + \frac{1}{3} + \frac{1}{4} + \frac{1}{5} + \frac{1}{6} + \frac{1}{7} + \frac{1}{8} + \ldots$$

$$> 1 + \frac{1}{2} + \left(\frac{1}{4} + \frac{1}{4}\right) + \left(\frac{1}{8} + \frac{1}{8} + \frac{1}{8} + \frac{1}{8}\right) + \ldots$$

$$> 1 + \frac{1}{2} + \frac{1}{2} + \frac{1}{2} + \ldots,$$

where we continuously replace each new subsequence of terms with length 2^k (where $k \geq 1$) in the original series with a smaller subsequence that sums to $\frac{1}{2}$. Thus, a *lower* bound on the sum, is infinity, and so $\lim_{n \to \infty} S(n) = \infty$.

3. George Gamow, *Matter, Earth, and Sky* (2nd ed.), Prentice-Hall, 1965, p. 20. Gamow didn't actually derive $S(n)$, as done here, but simply alluded to it.

4. This huge numerical value (it's *far* bigger than the number of stars in the Universe, estimated to be a "mere" 10^{22}) obviously can't be found by simply running a computer summation of the harmonic series. For an explanation of how it was computed, see R. P. Boas, Jr, and J. W. Wrench, Jr, "Partial Sums of the Harmonic Series," *American Mathematical Monthly*, October 1971, pp. 864–870, which gives the exact value of n for which $S(n)$ first exceeds 50: $n = 15092688622113788323693563264538101449859498$. Do you know how to even *say that*? I don't!

5. Leonard Eisner, "Leaning Tower of *the Physical Reviews*," *American Journal of Physics*, February 1959, pp. 121–122.

6. This discussion on dominos is inspired by a brief note written by Lorne A. Whitehead, "Domino 'chain reaction,' " *American Journal of Physics*, February 1983, p. 182.

12. Communication Satellite Physics

I [like] to talk about space to nonscientific audiences. In the
first place, they can't check up on whether what you are
saying is right or not. And in the second place, they can't
make head or tail out of what you are telling them
anyway—so they just gasp with surprise and wonderment,
and give you a big hand for being smart enough to say such
incomprehensible things. And I never let on that all you
have to do to work the whole thing out is to set the
centrifugal force equal to the gravitational force and solve
for the velocity [of the satellite]. That's all there is to it!
— *Lee A. DuBridge (president of Caltech), in a dinner talk at the
1960 spring meeting of the American Physical Society*

We rarely think of them—balls of metal crammed-full of densely packed electronics and bristling with antennas like porcupines hurtling around the Earth at speeds measured in miles per second while hundreds, even thousands, of miles over our head. And yet, every time we make a telephone call, watch a live television news show from Europe or the Middle East, or google something on the Internet, a communications satellite is almost surely involved somewhere. In this chapter I'll show you what DuBridge was talking about, with three calculations on the simple physics of these amazing creations of modern science, objects that were "crazy science fiction" little more than a few decades ago.

For our first calculation, let's go back to 1957, the year the Soviet Union launched the world's first satellite (*Sputnik 1*) into what is called a *low Earth orbit*. Sputnik zoomed above the surface of the planet at an altitude that varied from 132 miles to 582 miles, completing a full revolution every 96.2 minutes (called the *period* of the satellite). This value is a direct consequence of Newton's inverse square law of gravitation, and I show next how to derive the period.

While the orbit was elliptical, and not circular, we'll treat it as circular and justify that approximation as follows. Since the radius of the Earth is 6,380 kilometers, or 3,965 miles, *Sputnik*'s distance from the center of the Earth varied from 4,097 miles to 4,547 miles. That is, the distance was $4,322 \pm 225$ miles, or $4,322 \pm 5\%$ miles (which is $6,954 \pm 5\%$ kilometers). We'll assume as a crude first approximation that we can ignore that 5% variation and treat the orbit as circular with a radius of $R_s = 6.954 \times 10^6$ meters.

Now, let m and M be the mass of *Sputnik 1* and of the Earth, respectively. Following DuBridge, we set the gravitational acceleration of *Sputnik* equal to its centripetal acceleration and so write

$$\frac{\frac{GMm}{R_s^2}}{m} = \frac{v^2}{R_s},$$

where G is the universal gravitational constant we first encountered in Chapter 5, and v is the orbital speed of *Sputnik*. Thus,

$$v = \sqrt{\frac{GM}{R_s}}.$$

The period is then given by

$$T = \frac{2\pi R_s}{v} = 2\pi R_s \sqrt{\frac{R_s}{GM}}.$$

So, using $G = 6.67 \times 10^{-11} \frac{\text{meters cubed}}{\text{kilograms·seconds-squared}}$ and $M = 5.98 \times 10^{24}$ kilograms, we have

$$T = 2\pi (6.954 \times 10^6 \text{ meters})$$

$$\times \sqrt{\frac{6.954 \times 10^6 \text{ meters}}{\left(6.67 \times 10^{-11} \frac{\text{meters cubed}}{\text{kilograms·seconds-squared}}\right)\left(5.98 \times 10^{24} \text{ kilograms}\right)}}$$

$$= 43.693 \times 10^6 \sqrt{0.174 \times 10^{-7}} \text{ seconds}$$

$$= 43.693 \times 10^6 \sqrt{174 \times 10^{-10}} \text{ seconds}$$

$$= 576 \times 10^6 \times 10^{-5} \text{ seconds} = 5,760 \text{ seconds}$$

$$= 96 \text{ minutes},$$

which agrees quite nicely with the actual observed period of *Sputnik 1*.

Low Earth orbit is not a good orbit for a communications satellite; *Sputnik 1*, for example, was not visible from any point on Earth beneath its orbit for very long as it periodically zipped overhead from horizon to horizon. Each time the line of sight was broken, there was no way to communicate with the satellite until its return on the next overhead pass. Much more useful for communication is a satellite that remains fixed overhead, that appears to *hover* in the sky. That happens if the satellite is so high that its orbital period exactly matches (is *synchronized* to) the rotation period of the Earth. Such a satellite is said to be in *geosynchronous orbit*. How high is that orbit?

To answer that question, we return to the satellite period equation, which we now solve for R_s as a function of T. So,

$$T^2 = 4\pi^2 R_s^2 \frac{R_s}{GM} = 4\pi^2 \frac{R_s^3}{GM},$$

and so

$$R_s = \left(\frac{T^2 GM}{4\pi^2} \right)^{1/3}.$$

Setting $T = 86,400$ seconds, since the period of a geosynchronous satellite in orbit above the equator is one day (by definition!), we have

$$R_s = \left[\frac{\left[(86,400^2 \text{ seconds squared}) \left(6.67 \times 10^{-11} \frac{\text{meters cubed}}{\text{kilograms·seconds-squared}} \right) \right. }{4\pi^2} \right.$$
$$\left. \left. \times (5.98 \times 10^{24} \text{ kilograms}) \right] \right]^{\frac{1}{3}}$$

$$= \left[\frac{\left(8.64 \times 10^4\right)^2 \left(6.67 \times 10^{-11}\right) \left(5.98 \times 10^{24}\right)}{4\pi^2} \right]^{1/3} \text{meters}$$

$$= (75.42)^{1/3} \times 10^7 \text{ meters} = 4.225 \times 10^7 \text{ meters} = 42{,}250{,}000 \text{ meters}$$

$$= 26{,}258 \text{ miles.}$$

This is the distance from the *center of the Earth*, and so the *altitude* of a geosynchronous satellite above the *surface of the Earth* is

$$(26{,}258 - 3{,}965) \text{ miles} = 22{,}293 \text{ miles.}$$

There is another, clever way to calculate this answer. First, imagine a geosynchronous satellite in orbit, and then consider that it is not the only satellite the Earth has; there is also the Moon. Next, recall Kepler's third law from Chapter 5, which states that for a massive body (in Chapter 5 it was the Sun, and now it is the Earth) with satellites at various distances from the body's center, the square of the orbital period of each satellite is proportional to the cube of its average distance from the massive body. We know the Moon is 239,000 miles from the Earth (center to center) with an observed orbital period of 27.3 days. Our geosynchronous satellite has a period of one day and is a distance h from the center of the Earth. So, Kepler tells us that

$$\frac{(27.3)^2}{1^2} = \frac{(239{,}000)^3}{h^3} = 745.29$$

with h in miles. Thus,

$$h = \left(\frac{239{,}000^3}{745.29} \right)^{1/3} = \frac{239{,}000}{9.066} \text{ miles} = 26{,}362 \text{ miles}$$

from the *center of the Earth*. Thus, the *altitude* of a geosynchronous satellite above the *surface of the Earth* is

$$(26{,}326 - 3{,}965) \text{ miles} = 22{,}361 \text{ miles,}$$

which is pretty close to the result of our first calculation of the altitude.

A geosynchronous satellite is so high that there is essentially zero atmospheric drag on it, and the orbit is stable. That is not the case, however, for a low Earth orbit, where the satellite suffers significant atmospheric drag. *Sputnik 1*'s orbit, for example, decayed in just 3 months, and it fell back to Earth as a fireball. One surprising consequence of drag, one that is counterintuitive for most people, is that atmospheric *drag* on a satellite actually *increases* the speed of the satellite. We normally think of a drag force as a *retarding* or *slowing* force, but that is not so for satellites. This effect is startling enough that it is often called the *satellite paradox*.

To see how this comes about, let's write the drag force as f_d. Interestingly, we'll not need to know any of the details of f_d other than it is a *positive*-valued function (of the orbital speed, of the cross-sectional area of the satellite, and of the density of the atmosphere at orbital altitude). We'll start by calculating the total energy of the satellite, that is, the sum of its potential and kinetic energies (P.E. + K.E.). We assume that the center of the Earth in some coordinate system is at $r = 0$, and that the satellite is at distance $r = R_s$ from the center. Taking the zero of P.E. at infinity (this is the standard zero reference point used by physicists in astronomical analyses), and writing F as the gravitational force of Earth on the satellite, we have

$$\text{P.E.} = \int_{\infty}^{R_s} F\, dr = \int_{\infty}^{R_s} \frac{GMm}{r^2}\, dr = GMm$$

$$\times \int_{\infty}^{R_s} \frac{dr}{r^2} = GMm \left(-\frac{1}{r} \right) \Big|_{\infty}^{R_s} = -\frac{GMm}{R_s}.$$

The K.E. of the satellite is

$$\text{K.E.} = \frac{1}{2} m v^2,$$

where v is the orbital speed. As we showed earlier,

$$v = \sqrt{\frac{GM}{R_s}},$$

and so

$$v^2 = \frac{GM}{R_s}.$$

Thus,

$$\text{K.E.} = \frac{1}{2}\frac{GMm}{R_s},$$

and so the total energy is therefore

$$E = -\frac{GMm}{R_s} + \frac{1}{2}\frac{GMm}{R_s} = -\frac{GMm}{2R_s},$$

or, using the preceding boxed equation,

$$E = -\frac{1}{2}mv^2.$$

The atmospheric drag experienced by the satellite is an energy-loss mechanism, and the rate of energy lost by the satellite (the *power* dissipated) is given by vf_d (refer to note 2 in Chapter 3). That is,

$$\frac{dE}{dt} = -vf_d$$

where we insert the minus sign because we know that $vf_d > 0$ and that the total energy is *decreasing*. Now, from the boxed equation for total energy we have

$$v^2 = -2\frac{E}{m},$$

and so differentiation with respect to time gives

$$2v\frac{dv}{dt} = -\frac{2}{m}\frac{dE}{dt},$$

or

$$\frac{dv}{dt} = -\frac{1}{mv}\frac{dE}{dt}.$$

Using the boxed equation for $\frac{dE}{dt}$, we get

$$\frac{dv}{dt} = -\frac{1}{mv}\left(-vf_d\right) = \frac{f_d}{m},$$

which says the rate of change of the orbital speed of the satellite is directly proportional to the drag force. And since both f_d and m are positive, then $\frac{dv}{dt} > 0$, and the orbital speed continually *increases* even as the drag force continually *drains* energy from the satellite.

13. Walking a Ladder Upright

He had a dream in which he saw a stairway resting on the
earth, with its top reaching to heaven, and the angels of
God were ascending and descending on it.
— *Genesis 28:12*

In the Bible, Jacob only had to *dream* of an enormously long ladder joining Heaven and Earth upon which the angels could travel up and down between the two places (so why the wings? is a question beyond the ability of physics to answer). Actually erecting an even much shorter ladder, however, is not an easy task, as the following analysis shows.

A problem that virtually every homeowner eventually faces is that of raising a ladder to get access to the roof area, if only to retrieve the family cat, to remove a dead bird from the chimney, or to clean the gutters. A roof ladder is a cumbersome, slender object, one that is both long (20 or 30 feet) and fairly heavy, perhaps weighing up to 50 pounds or more. If we imagine such a ladder initially lying flat on the ground, how do we get it vertically upright without losing control of it and hurting ourselves or damaging a nearby structure? If the mathematician G. H. Hardy (see note 13 in the preface) ever had reason to think about the problem of raising a ladder to climb up to a roof—an event I would be willing to bet *never* occurred in Hardy's cloistered life!—I think he would have reconsidered his disparaging remark about the value of physics to the common man.

One method (used countless times by me) is to first drag the ladder over to the house and place the bottom end near a wall with the ladder at a right angle to the house. Then, going over to the far end of the

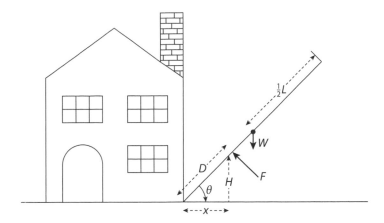

Figure 13.1. The geometry of raising a ladder

ladder, you pick it up and "walk the ladder up" as you approach the house. Pretty simple, right? Well, this seemingly innocuous technique has some hidden surprises for anyone who is doing it for the first time, ones that we'll discover as we apply some simple physics.[1]

Figure 13.1 shows a ladder of length L that is at angle θ to the ground. When the ladder is flat on the ground $\theta = 0°$, and when the ladder is fully raised $\theta = 90°(\frac{\pi}{2}$ radians$)$. A person walking the ladder up is at distance x from the bottom end of the ladder and is applying force F (perpendicular to) the ladder at a *constant* height H above the ground (H is the shoulder height of the person). This force is distance D from the bottom end of the ladder *as measured along the ladder*.

If we imagine the person raising the ladder by slowly reducing x ("walking the ladder in"), the situation shown in Figure 13.1 will be an equilibrium position if the clockwise rotation tendency of the ladder's weight is exactly balanced by the counterclockwise rotation tendency of the applied force F. Assuming the ladder has a weight W uniformly distributed over its length, and so is directed straight down from the point $\frac{1}{2}L$ along the length of the ladder, then the component of the weight W that is perpendicular to the ladder is $W\cos(\theta)$, and so the clockwise torque is $\frac{1}{2}WL\cos(\theta)$. Since the counterclockwise torque is FD, we have (for equilibrium)

$$FD = \frac{1}{2}WL\cos(\theta).$$

So,

$$F = \frac{WL\cos(\theta)}{2D},$$

or, since

$$H = D\sin(\theta),$$

then

$$D = \frac{H}{\sin(\theta)},$$

and we arrive at

$$F = \frac{WL\sin(\theta)\cos(\theta)}{2H}.$$

Since we have the trigonometric identity

$$\sin(2\theta) = 2\sin(\theta)\cos(\theta),$$

then

$$F = \frac{WL\sin(2\theta)}{4H}.$$

There is a *lot* of information in this last result. Remember, the only *variable* is θ, as W, L, and H are all constants. So, first of all, since $\sin(2\theta)$ is a constantly increasing function as 2θ varies from $0°$ to $90°$, we see that F will always be a maximum at $\theta = 45°$ for all W, L, and H, and the maximum force[2] is

$$F_{\max} = \frac{WL}{4H}.$$

If, for example, the ladder is 30 feet long and weighs 50 pounds, then a person with a shoulder height of 5 feet will have to be prepared to exert a force of

$$\frac{(30)(50)}{4(5)} \text{ pounds} = 75\,\text{pounds}$$

when the ladder is tilted at $45°$ (that is, when the person is distance $x = H = 5$ feet from the bottom of the ladder). This force is greater than the weight of the ladder, a result that almost always surprises.

The radio amateur I mentioned earlier had a 60-foot antenna tower weighing 120 pounds to raise, and with a 5-foot shoulder height, he had to exert a maximum force (at $\theta = 45°$, and so when just 5 feet from the bottom of the tower) of

$$\frac{(60)(120)}{4(5)} \text{ pounds} = 360 \text{ pounds},$$

three times the tower weight.

Another interesting exercise is to calculate the force required as a function of how far a person is from the bottom of the ladder, that is, of F as a function of x. We have

$$\tan(\theta) = \frac{H}{x},$$

and so

$$\theta = \tan^{-1}\left(\frac{H}{x}\right).$$

Thus,

$$F = \frac{WL\sin\left\{2\tan^{-1}\left(\frac{H}{x}\right)\right\}}{4H}, \quad 0 \leq x \leq L.$$

For given values of W, L, and H it is easy to plot F as x varies, and Figure 13.2 shows the result for the radio amateur raising his antenna tower ($H = 5$ feet, $L = 60$ feet, and $W = 120$ pounds).

As the radio amateur wrote at the end of his paper, the real surprise in this problem is now made clear by Figure 13.2: "This curve shows that the maximum force occurs after you have walked [55 feet]! At this point, if the load seems unbearable, you are faced with a long walk back should you decide you can't handle the tower. Here is where most accidents occur. Even if you can lift 360 lb, you must remember that you have already walked quite a distance supporting over 100 lb." This caution is an important one for all homeowners to keep in mind when thinking of getting on their roof.

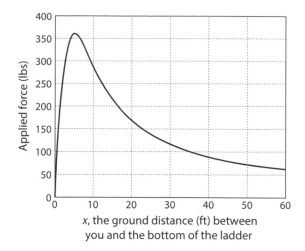

Figure 13.2. Raising a 120-pound, 60-foot tower

Notes

1. This problem was motivated by my reading of the task a radio amateur faced in raising an antenna tower that was 60 feet long and weighed 120 pounds; see P. B. Mathewson, "Walking Your Tower Up? Can You Do It Safely?" *QST*, March 1980, pp. 32–33. Three years later, in September 1983, an article arriving at the same results appeared in *The Physics Teacher* ("Practical Mechanics: Raising a Mast," pp. 379–380) by Robert L. Neman.

2. In the paper that repeats the results of the earlier paper in *QST*, Neman (see note 1) introduces a needless complication by missing the trigonometric simplification used here, namely, $\sin(2\theta) = 2\sin(\theta)\cos(\theta)$. Rather, he finds the maximum force by performing a differentiation. As a *Physics Teacher* reader from Denmark observed (September 1984, p. 350), "The rather lengthy derivation used [by Neman] ... is unnecessary." A simple physics problem should be *kept* simple, and this is another good example of the moral in the story I related concerning Edison and mathematics at the end of Chapter 1: "Don't use a cannon (calculus) when a peashooter (high school trigonometry) will do the job."

14. Why Is the Sky Dark at Night?

There are no paradoxes in science.
—*Lord Kelvin*, The Baltimore Lectures[1]

This chapter is on a topic that demonstrates how what *appears* to be among the most benign, commonplace, and indeed just plain ordinary observation is actually among the most profound questions that physicists have ever asked. So, let's jump right in with a centuries-old question, one that at first may appear to be ridiculous (or, at least, metaphysical): Why is the night sky dark? (Try this on a friend, even one trained in science, and don't be surprised to hear the answer, "Of course it's dark, you idiot, it's *night*!"). It took a genius to appreciate that it is *not* a ridiculous question.[2]

After all, if space is infinite, containing an infinite number of uniformly distributed stars, then every line of sight as you look out into space should eventually intercept the surface of a star (indeed, probability theory *demands* it, as I'll argue in just a bit), as indicated in Figure 14.1. The night sky shouldn't be dark at all but, instead, blindingly bright. But it's not bright. Why not?

An easy way out of this puzzle would seem to be simply to deny that space (and the number of stars in it) is infinite. But that would be giving up a lot (no pun intended). An infinite space avoids the embarrassing question of what is "beyond" the end of a universe of finite size. Early theologians, in particular, liked an infinite space because it avoided the issue of God having any limitations on His abilities, and they made time infinite, too, to avoid the equally embarrassing question of what God was doing before He created everything a finite time ago.

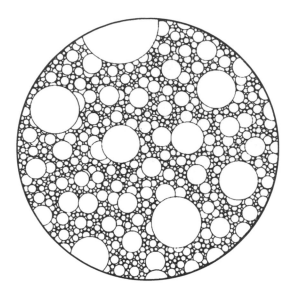

Figure 14.1. A star, everywhere you look

(Wits famously replied to that with "creating Hell for all who ask.")
Modern theologians, some with PhDs in theoretical physics, are more
sophisticated about these questions.

So, let's assume all the *infinite* stars are just as bright as is our
Sun which, even with a surface temperature of 11,000 °F, is quite an
ordinary star. The night sky should then be intensely bright because
of a very simple geometric argument, based on Figure 14.2. In fact, by
the argument I'm about to show you, the night sky should be *infinitely*
bright, and all space should be flooded with a radiation level that
would instantly vaporize the Earth and everything (including us) on it.

Imagine yourself as the observer in Figure 14.2, surrounded by
infinite space, a space containing infinitely many, uniformly distributed stars. Let this space be divided into concentric spherical shells
(the figure shows just two of those shells), with each shell having
the same depth (thickness) of ΔR. The volume of the shell that is
distance R from the observer is approximately $4\pi R^2 \Delta R$ (an excellent
approximation for $\Delta R \ll R$), and this volume is a direct measure of
the number of stars (the asterisks in Figure 14.2) in the shell. The light
intensity you see from a single star at distance R varies as $\frac{1}{R^2}$, and so

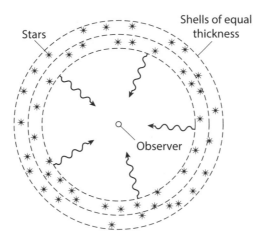

Figure 14.2. Two spherical shells of stars

the light intensity you see from *all* the stars in the shell at distance R varies as $\left(4\pi R^2 \Delta R\right) \frac{1}{R^2} = 4\pi \Delta R$. But this value is a *constant*; that is, the light intensity from a shell is independent of the shell's distance from the observer and depends only on its thickness. So, since there is an infinity of shells, the observer should "see" an infinite light intensity.[3] Well, we don't see anything even remotely like that, and so we appear to have a paradox. But Lord Kelvin denied such a thing can occur, and so there must be a resolution. I'll let you think about this problem for a while and then tell you more at the end of the chapter.

To see the probability connection I mentioned earlier behind the night sky question, imagine the usual x, y-coordinate system. If we randomly draw a straight line in the lower half-sector of the first quadrant of this system (the line makes a polar angle with the x-axis in the interval $0 < \theta < 45°$), starting from the origin and extended out to infinity, what is the probability that the line passes through at least one lattice point other than the origin? (A lattice point is a point with integer coordinates, and so $(3, 7)$ is a lattice point, but $(\pi, \sqrt{2})$ is not.) For this question, the definitions of a line and a point are purely mathematical. That is, a point has no size (extension) in any direction, and a line has zero width. So, again, what's the probability that the line passes through at least one lattice point other than the origin?

The probability is zero.[4] Here's why. If the line does pass through the lattice point (x_k, y_k), then the line makes angle θ with the x-axis, where $\tan(\theta) = \frac{x_k}{y_k}$, which is a rational number in the interval 0 to 1. But the rationals are a *countable* infinity (that is, they can be put into a one-to-one correspondence with the positive integers), while all possible values for $\tan(\theta)$ are the real numbers in the interval 0 to 1, which are an *uncountable* infinity.[5] It is important to note, however, that if around each lattice point we draw a circle of radius ε (think of the circular cross sections of the stars in the night sky question), then the probability that a randomly drawn line passes through an *infinite* number of circles is 1 *no matter how small* ε may be, as long as $\varepsilon > 0$. (This is not trivial to prove!)

This imagery seems to be the key to avoiding the "infinite intensity" conclusion. Unfortunately, as you'll see, we'll succeed only in reducing the night sky brilliance from infinity to "just" that of the surface of a star! That is a really big drop, yes, but not enough to save us from being cremated in a cosmic furnace. Instead of being turned into toast in 10^{-30} seconds, it will now take a trillion times longer, or 10^{-18} seconds. Although this period is still on the brief side, there's precious little comfort for us there. Nevertheless, let's first see how to go from $\infty\,°F$ to $11,000\,°F$. It's quite simple: the idea is to invoke what is called the *lookout distance*. (Later in this chapter I'll show you how to calculate this distance.) That is how far your line of sight extends until it intercepts the surface of a star. That star then blocks all the stars behind it, and so you *don't* receive radiation from *infinitely* many stars. Not to put too grim a spin on this, however, $11,000\,°F$ is still pretty hot.

Actually, the "blockage" idea doesn't work, either. It was endorsed by both Olbers and Chéseaux, but they wrote before the principle of the conservation of energy entered physics (in the 1840s). That principle is the fatal flaw in the blockage idea, because any interstellar matter that absorbed the light energy of more distant stars would then experience a temperature rise and therefore would simply reradiate that absorbed energy toward us. And, even if blockage *did* work, it wouldn't do any good anyway. Here's why.

In an end-of-chapter note in his book, Harrison (note 2) gives a nice, elementary math analysis of the implications of what a "solid sky of stars" means, even if there are no more stars behind the ones we see. I can't think of any way to improve on his presentation and so here it is, just as Harrison wrote: "The sky has an angular area of

4π square radians (a square radian is a unit of solid angle, the steradian). A radian equals $180/\pi = 57.3$ arc degrees, and [so] the whole sky is covered with $4 \times 180^2/\pi = 41{,}253$ square degrees. The Sun subtends an angular radius of almost 0.27 degrees, corresponding to an area little more than 0.22 square degrees. The area of the whole sky is therefore roughly 180,000 times that of the Sun. In other words, a bright-sky universe pours down on Earth 180,000 times as much radiation as the Sun." That isn't infinity, but it's still large enough to vaporize the Earth. (This result was first calculated by Chéseaux; see note 3.)

So, what *is* the answer? In an astonishing bit of historical research, Harrison traced the basic idea of the modern answer to, of all people, the American poet Edgar Allen Poe (1809–1849)!—specifically, to his long (well over 100 pages) essay *Eureka: A Prose Poem*, published in 1848. Poe's idea was simply that the universe is so vast that there is a distance beyond which the stars are so remote that their light hasn't yet had enough time, since their creation, to reach Earth. That distance is a horizon that marks the ends of the visible universe, a horizon that recedes from Earth at the speed of light. This simple idea instantly escapes the disaster of an *infinitely* bright night sky but, however, it does allow the night sky to grow ever brighter (hotter) as ever more stars become visible.

Poe's idea *is* part of the answer to the night sky question, but there was far too much talk in *Eureka* about God to encourage scientists to take it seriously. And there was too much arithmetic in it to encourage nonscientists to wade through all the numbers Poe endlessly paraded in front of the reader to show how vast is the universe. A comment by one analytical reader—Irving Stringham (1847–1909), a professor of mathematics at the University of California–Berkeley—gives a good sense of how the scientific community viewed *Eureka*: "Poe believed himself to be that extinct being, a universal genius of the highest order; and he wrote this essay to prove his powers in philosophy and in science. . . . Poe succeeded only in showing how egregiously genius may mistake its realm."[6] In other words, Poe should have stuck with poems and short stories, and left astronomy to the astronomers. I think this assessment a bit too harsh (*Eureka* is, in my opinion, a fascinating read), but it does illustrate the general reaction to Poe by many scientists.

Poe wasn't the only one to have the idea that the sheer size of the universe is the key to the answer to the dark night sky puzzle. The American theoretical physicist Frank Tipler, for example, has suggested that the answer to the "dark sky at night" question was solved in 1861 by the German astronomer Johann Heinrich von Mädler (1794–1874). (Poe had clearly read Mädler's earlier writings, as he specifically mentions him several times in *Eureka*.) In the fifth edition of his book *Popular Astronomy*, Mädler wrote "[Since] the velocity of light is *finite*, a *finite* amount of time has passed from the beginning of Creation until our day, and we, therefore, can only perceive the heavenly bodies out to the distance that light has travelled during that finite time. As the dark background of the heavens is sufficiently explained in this manner, indeed presents itself as necessary, the compulsion to assume light [blockage] is eliminated. Instead of saying the light from those [blocked] distances does *not* reach us, one must say: it has *not yet* reached us."[7]

Poe and Mädler were right, as far as they went, but they didn't go far enough. In another impressive bit of scholarly digging, Harrison uncovered a long-forgotten paper (that apparently made little impression even when new) by Lord Kelvin, published in 1901 (reproduced in Harrison's book). There we see the final part to the answer to the night sky problem—stars don't shine forever but, rather, have a finite lifetime. So, when we do see the light from a star, it's only for a finite duration. Kelvin's reputation was immense in Victorian times, in part because of his famous calculation of the age of the Sun. His estimate (no more than 500 million years, and most probably as short as 50 million years) was far too brief, because he knew nothing of the nuclear reactions that power the Sun.[8] The specific number is not important, however, only that it is finite. Today we believe the Sun is about five billion years old, and has about the same time left to shine. The important concept, however, is that while 10 billion years is "long," it's finite, and Kelvin's earlier work had made that an established fact (based on what he called "irrefragable dynamics") in his mind.

To understand what Kelvin's 1901 paper proposed, imagine for the sake of argument an infinite space, filled with an infinity of uniformly distributed stars, surrounding the Earth. Imagine that all these stars were "turned on" at the same time by Poe's God. Light from the nearest stars would "soon" arrive at Earth and would eventually be joined by the light from even more distant stars. After 10 billion years or so, however, those nearest stars would cease to shine, and an expanding sphere (centered on the Earth) of dark stars would start to appear. Then, however, light from beyond 10 billion light-years would start arriving at Earth to replace the light lost from the stars going dark. In this way, the total light arriving at Earth would reach a steady-state equilibrium in the total starlight at night. How bright a night sky would that equilibrium be? Not bright at all, according to Kelvin. Here's how he calculated that brightness, using just geometry, a touch of algebra, and an easy integration.

We assume all stars are the same size, with radius a, and that they are randomly (uniformly) distributed throughout space with an average density of n stars per unit volume. Then, centered on the Earth, we construct a spherical shell of radius q and thickness dq. The number of stars in this shell is the volume of the shell times n, that is, $4\pi q^2 dq\, n$. The total area of the shell surface that is covered by the cross-sectional areas of these stars is

$$\left(\pi a^2\right)\left(4\pi q^2 dq\, n\right) = 4\pi^2 n a^2 q^2\, dq\,.$$

Dividing this covered area by the total area of the shell we obtain the fraction f of the more distant sky that is blocked from view by the stars in the shell:

$$f = \frac{4\pi^2 n a^2 q^2 dq}{4\pi q^2} = \pi n a^2\, dq\,.$$

We write $\sigma = \pi a^2$ as the cross-sectional area of a star, and so

$$f = n\sigma\, dq\,.$$

If we let q vary from 0 to some value r, then the *total* fraction of the sky blocked from view by all the nested shells inside a sphere of radius r is

$$\int_0^r f\, dq = \int_0^r n\sigma\, dq = n\sigma r = \frac{r}{\lambda}\,, \quad \lambda = \frac{1}{n\sigma}\,,$$

where λ is the lookout distance mentioned earlier. In this calculation I'm ignoring (as Kelvin himself explicitly acknowledged) the eclipsing of distant stars by nearer stars; he claimed such a masking event would be "extremely rare."

To evaluate λ, we need to know n. Suppose that there are N stars in a sphere of radius r.

Then,

$$n = \frac{N}{\frac{4}{3}\pi r^3} = \frac{3N}{4\pi r^3},$$

and so,

$$n\sigma r = \left(\frac{3N}{4\pi r^3}\right)\left(\pi a^2\right)r = \frac{3N}{4}\left(\frac{a}{r}\right)^2,$$

the fraction of the sky covered by the N stars. When Kelvin wrote, he subscribed to the general turn-of-the-twentieth-century view that the Milky Way galaxy, alone, *was* the universe. It wasn't until after his death that the modern view of a universe with 10^{11} galaxies, each with 10^{11} stars (for a total of 10^{22} stars!), was developed. For Kelvin, there was just the Milky Way galaxy with 10^9 stars, all contained in a sphere with radius 3.09×10^{16} kilometers (3,300 light-years), giving a density of

$$n = \frac{3 \times 10^9}{4\pi \left(3.3 \times 10^3\right)^3} \text{ stars/cubic light-year}$$

$$= 0.0066 \text{ stars/cubic light-year},$$

that is, one star per 150 cubic light-years, on average.

This may, at first glance, seem to be a pretty thin distribution density, but a second look may give you reason to reconsider. This density is equivalent to 10 stars randomly sprinkled throughout 1,500 cubic light-years or, in other words, throughout the interior of a sphere of radius 7.1 light-years. Now, an intuitively satisfying way to measure "how close" these stars are to one another is to look at the average value of the *nearest-neighbor distance*. That is, for each of the 10 stars, how far away on average is the nearest star? (Note, carefully, that the nearest-neighbor function is not a reciprocal one. That is, if A's closest neighbor is B, B's closest neighbor is not necessarily A.) This is a problem in probability, one that can be solved exactly *if* one has just a bit more math than I am assuming here, and so I'll simply tell you the answer:[9] if we take one of the stars as the center of a sphere of radius r, and scatter the other nine stars at random throughout the sphere, then the average nearest-neighbor distance is $0.4191r$ ($= 3$ light-years for $r = 7.1$ light-years). For comparison, the Sun's nearest stellar neighbor is the red dwarf Proxima Centauri, part of the Alpha Centauri triple-star system, 4.3 light-years distant.

Now, the radius of the Sun is 7×10^5 kilometers or, converting to light-years (using 3×10^8 meters/second as the speed of light),

$$a = \frac{7 \times 10^8 \text{ meters}}{3 \times 10^8 \frac{\text{meters}}{\text{second}} \times 3,600 \frac{\text{seconds}}{\text{hour}} \times 24 \frac{\text{hours}}{\text{day}} \times 365 \frac{\text{days}}{\text{year}}}$$

$$= 7.4 \times 10^{-8} \text{ light-years.}$$

So, the Sun's cross-sectional area is

$$\sigma = \pi \left(7.4 \times 10^{-8}\right)^2 \text{ light-years squared}$$

$$= 172 \times 10^{-16} \text{ light-years squared,}$$

which gives a lookout distance in Kelvin's universe of

$$\lambda = \frac{1}{(0.0066)\left(172 \times 10^{-16}\right)} \text{ light-years} = 8.8 \times 10^{15} \text{ light-years.}$$

In other words, when you look out into the night sky of Kelvin's assumed universe, your line of sight has to extend out nearly 9 quadrillion light-years to terminate on the surface of a star. More dramatically (if that's possible), you'd be seeing light that left a star 9 quadrillion years ago—*but the universe isn't that old*, so you in fact see nothing, and the night sky is (on average) dark.

To really drive that conclusion home, if we evaluate Kelvin's expression for the fraction of the sky covered by N stars ($n\sigma r$), we get

$$n\sigma r = \frac{3N}{4}\left(\frac{a}{r}\right)^2 = \frac{3 \times 10^9}{4}\left(\frac{7.4 \times 10^{-8}}{3.3 \times 10^3}\right)^2 = 3.8 \times 10^{-13},$$

which really *is* pretty darn small! One can, of course play with the values of N and r—today we think N should be a lot bigger than 10^9, but we also think r should be a lot bigger than 3,300 light-years—with the final result appearing to be quite insensitive to such number games. As Kelvin himself concluded, "it seems there is no possibility of having enough stars ... to make a total of star-disc-area more than 10^{-12} or 10^{-11} of the whole sky." So, the next time your significant other snuggles up close and comments on how romantic the dark night sky is with its sprinkle of stars, you can now reply, "Gee, do you know *why* it's mostly dark? Why there aren't stars *everywhere* you look? Let me tell you the story behind that. It's all because ... "

See how well *that* works out for you!

Notes

1. *The Baltimore Lectures* are a stenographic record of a series of lectures delivered by Scottish Professor William Thomson (1824–1907), also called Lord Kelvin, in October 1884 at The Johns Hopkins University.

2. This question is often (and erroneously) discussed under the name *Olbers' paradox*, after the German astronomer Heinrich Wilhelm Olbers (1758–1840), who wrote of it in 1823. In fact, it is due to Kepler (see Chapter 5), who had posed it more than two *centuries* (!) earlier, in 1610. It didn't appear in print, however, until Newton's friend Edmund Halley (1656–1742) discussed it (alas, incorrectly) in a 1772 paper (Olbers' 1823 paper pointed out Halley's error). An outstanding history of the dark night sky question is by Edward Harrison, *Darkness at Night*, Harvard University Press, 1987, which reproduces Halley's and Olbers' papers.

3. This argument is due to the Swiss astronomer Jean-Philippe Loys de Chéseaux (1718–1751), who presented it as an appendix to a 1744 book on comets (!), which is why it was probably not well known until many years after. This appendix is reproduced by Harrison in his book (note 2). A *faint* glimmer of the shell argument can be found in Halley.

4. A probability of zero does *not* mean a line passing through a lattice point is impossible, as you can obviously draw *infinitely* many such lines. There's just a "bigger infinity" of lines of which *none* passes through even one lattice point. An impossible event does indeed have probability 0, but the converse is *not* true.

5. You can find high school–level proofs of these statements in my book *The Logician and the Engineer*, Princeton University Press, 2013, pp. 168–173.

6. Quoted from *The Works of Edgar Allen Poe in Ten Volumes* (vol. 9), E. C. Stedman and G. E. Woodbury (eds.), The Colonial Company, 1903, p. 312.

7. Quoted from Frank J. Tipler, "Johann Mädler's Resolution of Olbers' Paradox," *Quarterly Journal of the Royal Astronomical Society*, September 1988, pp. 313–325.

8. All stars, including the Sun, are powered by nuclear fusion reactions deep in their interior. Since knowledge of those reactions lay far in Kelvin's future (indeed, after his death), he had to find some other mechanism for the stellar energy source. The only possible candidate in his day was the *gravitational contraction* of interstellar gas clouds. During contraction, the potential energy of the collapsing gas goes into increasing the kinetic energy of the gas molecules, thus heating the gas into radiance. Ironically, we believe today that gravitational contraction is indeed what *starts* star formation, heating a collapsing gas cloud up to the point where fusion reactions can start and so stop the collapse. Thus, Kelvin wasn't totally wrong. You can find some detailed calculations on what Kelvin did in my book *Mrs. Perkins's Electric Quilt*, Princeton University Press, 2009, pp. 157–162.

9. You can find a complete analysis in *Mrs. Perkins's* (note 8), pp. 285–298, 365–366.

15. How Some Things Float (or Don't)

Iron in the water shall float
As easy as a wooden boat.
— *prophecy attributed to Mother Shipton, an English witch*
(according to legend) who lived in seventeenth-century Yorkshire

To start this chapter on a somewhat less than scholarly note, consider the following short fable. Bob Bankrobber, master criminal, was recently caught by his Gangland Boss skimming money from bank heists before handing over the loot. That's why he is now standing in a boat floating in the middle of a large lake, with his feet up to the ankles stuck in a big bucket of hardening cement. Two of his soon-to-be-former nefarious colleagues, Fred Firebug and Tom Thug, have orders from the Boss to toss Bob overboard. Just before they are to do that, Fred says to Tom, "Hey, Tom, before I learned how to start fires I was a physics major at State U, and this reminds me of a homework problem I once had. After we toss Bob in, and he sinks to the bottom, will the water level in the lake rise or fall?"

Tom, who flunked out as a major at State U in Extreme Leisure Studies, thinks this over and quickly becomes confused. After all, when Bob enters the lake he will displace some water, which should cause the water level to rise. However, once Bob has left the boat it will float higher and so displace *less* water, which should cause the water level to fall. Which effect dominates?

Tom may not be overly bright but, for a thug, he's somewhat honest and so replies, "Gee, Fred, I don't know." Tom isn't totally stupid, though, and he gets a really good idea. "Let's measure the lake level *before* we toss Bob and once again *after* we toss him." He then pulls a piece of chalk out of his pants pocket and makes a mark at water level on a vertical pole that just happens to be right next to the boat, sticking up out of the lake with its other end buried in the lake bottom. "See, Fred," says Tom, "all we have to do is observe if the lake water level, after we toss Bob in, is above or below the chalk line." Fred sees the logic of that, and agrees that Tom's idea makes sense. Despite his predicament, even Bob (who was a math major at State U before yielding to the wicked temptations of the dark side) finds the question provocatively interesting and is about to add his thoughts when over the boat's edge he goes, and so whatever he was going to add to the discussion is lost to history.

So let's forget Bob and concentrate on Fred's question: will the water level rise or fall (or, perhaps, not change)?

Or, consider this variation: Tom and Fred, reluctant to "terminate with extreme prejudice" their old pal Bob, decide to give him at least a chance to survive while still obeying the letter of the Boss' order. They skip the cement and simply toss Bob into the lake, where he doesn't sink but, rather, *floats*. How does the water level change in that case?

Our solutions to these questions will be based on one of the oldest laws of physics, one known in antiquity: *Archimedes' principle*, discovered in the third century BC. According to the famous story, Archimedes (287?–212 BC) resolved a dilemma for King Hiero II of Syracuse, Sicily: was the king's royal crown made of pure gold, or had the goldsmith pocketed some of the gold and, to cover the theft, replaced the stolen gold with an equal weight of silver? As the story is told in countless physics textbooks, Archimedes suddenly realized how to answer that question while taking a bath, and so excited was he by his revelation that he leaped from the water and ran naked through the streets crying *Eureka!* Just *what* the great man's solution was remains a mystery, however, as he wrote nothing about it, and, in fact, the story of the king's crown wasn't first mentioned until two centuries later, in the Roman architect Marcus Vitruvius's book *On Architecture*.[1]

The principle is easy to state: an object floating or wholly submerged in a fluid experiences a buoyant force equal to the weight of the fluid displaced by the object. In the case of a totally submerged object, the *volume* of the fluid displaced is, of course, the *volume* of the object.

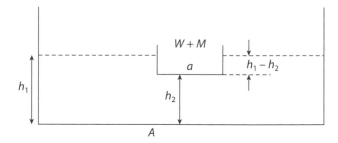

Figure 15.1. Before Bob and his cement go over the side of the boat

In physics textbooks the principle is usually illustrated by considering how fluid pressure varies with depth, and this allows the computation of the net upward force on the object (upward, or buoyant, as the pressure is greater on the bottom surfaces of the object than it is elsewhere).[2]

The solutions I'll show you here, to both our questions about the changes in water level, will use the principle in an *analytical* way. That is, I'll be writing some equations. Some commentators on these questions have written disparagingly about such an approach, preferring to argue their way to a conclusion with words alone. One writer I once came across expressed that position by saying something like, "Of course, a math-obsessed physicist will immediately attack such questions by writing down a whole bunch of equations and solving them."

That's *bad*? That's what you *have* to do when faced with questions a lot tougher than tossing Bob into the lake, problems for which you can't easily come up with a facile word solution. (Before we are through with this chapter I'll show you an example of an "Archimedes problem" in which the analytical approach is a *must*.) So, to show you how to be analytical with our two "let's toss Bob" simple physics questions, I'll write a few equations (hardly a "whole bunch"), and you'll see how smoothly, systematically, and *quickly* we arrive at the answers. I'll start with the first question.

Figure 15.1 illustrates Bob's situation just *before* Tom and Fred toss him and his cement-coated shoes overboard. The weight of the boat plus Tom and Fred is W, and the weight of Bob and his cement is

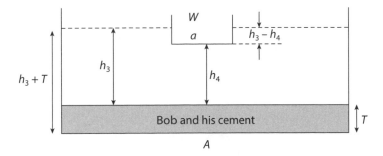

Figure 15.2. After Bob and his cement go over the side of the boat

M. We'll take the density of water as ρ_w, and the density of Bob and his cement as $\rho > \rho_w$ (because Bob and his cement *sink*). The cross-sectional area of the boat (assumed to have vertical sides) is a, and the cross-sectional area of the lake (assumed to have vertical walls) is A. The bottom of the lake is perfectly flat, and the water level of the lake is h_1. The bottom of the boat is h_2 above the bottom of the lake, where, of course, $h_1 > h_2$.

Now, this is, of course, all a pretty grim business for Bob, but as steely-eyed analytical physicists let's stop thinking of Bob as Bob but simply as a mass that displaces a quantity of water equal to the volume of Bob and his cement. The actual Bob will come to rest as a big blob on the lake bottom, but as far as displacing water, we can imagine Bob and his cement as uniformly spread over the entire lake bottom in a layer of thickness T. This is shown in Figure 15.2, which is the situation *after* Bob has paid the price for being a crooked crook.

Okay, let's get analytical. The first equation we can write immediately is that of "conservation of water"; that is, the amount of water in the lake is the same after Bob is tossed as it was before he was tossed. So,

$$(A - a)h_1 + ah_2 = (A - a)h_3 + ah_4. \tag{1}$$

Before Bob is tossed, the boat plus all three men weighs $W + M$, and because the boat and men are floating, we know from Archimedes' principle that they are displacing a volume of water that weighs $W + M$. Remembering that the density of water is ρ_w, this weight of water has volume $\frac{W+M}{\rho_w}$. But the displaced volume is clearly given by $a(h_1 - h_2)$, and so we have

$$a(h_1 - h_2) = \frac{W + M}{\rho_w}. \tag{2}$$

After Bob is tossed, the boat plus *two* men (just Fred and Tom as, alas, Bob is elsewhere) weighs W, and, again, as they are floating, the same reasoning we used to get (2) says

$$a(h_3 - h_4) = \frac{W}{\rho_w}. \tag{3}$$

Finally, since cemented Bob weighs M and has density ρ, his volume is $\frac{M}{\rho}$, which must equal AT (the volume of Bob and his cement when uniformly spread over the lake bottom). So,

$$T = \frac{M}{\rho A}. \tag{4}$$

Now, carefully keep in mind what we are after. We have *three* equations, (1), (2), and (3), in four unknowns, h_1, h_2, h_3, and h_4, plus the auxiliary, (4), which isn't an equation (it is simply a relation expressing T in terms of three *known* quantities). Since you need *four* equations to solve for four unknowns, you might think we are dead in the water (somewhat like Bob). But not so! That's because we want to know only the *change* in water level. Referring to Tom's chalk mark at height h_1 above the lake bottom before Bob is tossed, the new water level after Bob is tossed is $h_3 + T$, and so what we are after is $h_1 - (h_3 + T) = h_1 - h_3 - T$, which we'll then check to see whether it is negative (water level rose), zero (water level didn't change), or positive (water level fell). Since we seek only the *difference* of two unknowns (h_1 and h_3, because T is *not* an unknown), our three equations in four unknowns are sufficient to do the job. From this point on, it's all just simple algebra.

From (1),

$$h_1 + \frac{a}{A-a}h_2 = h_3 + \frac{a}{A-a}h_4,$$

or

$$h_1 - h_3 = \frac{a}{A-a}h_4 - \frac{a}{A-a}h_2. \tag{5}$$

From (2)

$$h_1 - h_2 = \frac{W+M}{a\rho_w},$$

and from (3)

$$h_3 - h_4 = \frac{W}{a\rho_w}.$$

So,

$$h_2 = h_1 - \frac{W+M}{a\rho_w},$$

and

$$h_4 = h_3 - \frac{W}{a\rho_w}.$$

Substituting these results for h_2 and h_4 into (5), we have

$$h_1 - h_3 = \frac{a}{A-a}\left(h_3 - \frac{W}{a\rho_w}\right) - \frac{a}{A-a}\left(h_1 - \frac{W+M}{a\rho_w}\right)$$

$$= \frac{a}{A-a}h_3 - \frac{W}{(A-a)\rho_w} - \frac{a}{A-a}h_1 + \frac{W+M}{(A-a)\rho_w}$$

$$= \frac{a}{A-a}h_3 - \frac{a}{A-a}h_1 + \frac{M}{(A-a)\rho_w}.$$

Therefore,

$$h_1 + \frac{a}{A-a}h_1 = h_3 + \frac{a}{A-a}h_3 + \frac{M}{(A-a)\rho_w},$$

or

$$h_1\left(1 + \frac{a}{A-a}\right) = h_3\left(1 + \frac{a}{A-a}\right) + \frac{M}{(A-a)\rho_w},$$

or

$$h_1 \frac{A}{A-a} = h_3 \frac{A}{A-a} + \frac{M}{(A-a)\rho_w},$$

or

$$h_1 A = h_3 A + \frac{M}{\rho_w},$$

and so,

$$h_1 - h_3 = \frac{M}{A\rho_w}.$$

Finally, using (4), we obtain

$$h_1 - h_3 - T = \frac{M}{A\rho_w} - \frac{M}{A\rho} = \frac{M}{A}\left(\frac{1}{\rho_w} - \frac{1}{\rho}\right) = \frac{M}{A}\left(\frac{\rho - \rho_w}{\rho\rho_w}\right) > 0,$$

because $\rho > \rho_w$. So, the answer to Fred's question is, the lake water level *drops* when Bob and his cement are tossed overboard. A virtue of the analytical approach is that if we are told the values of M, A, and ρ (we can look up ρ_w in a table of physical constants), we can calculate by *how much* the level drops.

You'll recall that at the start of this chapter I mentioned "facile word solutions" as an alternative to our analytical approach. Here's an example of such an argument. Suppose Bob and his cement are *very* dense, so dense in fact that, for a given weight M, there is hardly any volume to Bob and his cement. Thus, when Bob goes over the side, the boat will float higher (which causes the water level to fall), and yet Bob will displace hardly any water as he sinks to the bottom (and so his presence in the lake will have hardly any effect on the water level). The net effect will be that the water level falls, just as we derived. This is clever, but it is for an extreme case. How do we know it's *always* valid? The analytical approach avoids this issue and *proves* the water level *always* falls for all possible values of W, M, A, a, and $\rho > \rho_w$, and even tells us by how much the level falls.

Now, on to the second question: what if Bob *floats* because Fred and Tom skip the cement? We have, for the "after Bob is tossed" case, the situation shown in Figure 15.3. The unknowns h_1, h_2, h_3, and h_4 are as before, but now we have an additional one, h_5 (the distance Bob floats

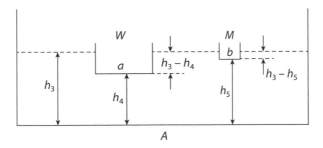

Figure 15.3. After Bob goes over the side of the boat and *floats*

above the bottom of the lake). Curiously, even though we have an extra unknown, many people find this case to be more "obvious" than the first one, reasoning as follows. Before Bob is tossed, he is floating (in the boat), and after he is tossed he is floating on his own. In both cases he is floating, and the lake doesn't "know" whether he is in a boat or not. So, the water level of the lake shouldn't change.

In fact, that conclusion is correct, as I'll show you analytically in just a moment. However, if you were to ask 100 people this question, I would wager there would be at least a few who would not be quite so sure. If you were to show these same 100 people an analytical solution, though, I would bet there wouldn't be *any* doubters. So, let's be analytical again. I'll assume that Bob floats with vertical sides and has cross-sectional area b, as shown in Figure 15.3.

From the conservation of water we have

$$(A - a)h_1 + ah_2 = ah_4 + bh_5 + (A - a - b)h_3. \tag{6}$$

Equation (2) still holds, and so

$$a(h_1 - h_2) = \frac{W + M}{\rho_w}. \tag{7}$$

And (3) also still holds, so

$$a(h_3 - h_4) = \frac{W}{\rho_w}. \tag{8}$$

Finally, writing the physics of a *floating* Bob,

$$(h_3 - h_5)b = \frac{M}{\rho_w}. \tag{9}$$

Thus, we have four equations in five unknowns, which are all we need to solve for $h_1 - h_3$, the *change* in the lake water level.

From (6),

$$(A - a)h_1 + ah_2 = ah_4 + bh_5 + (A - a)h_3 - bh_3,$$

or

$$(A - a)h_1 + ah_2 = ah_4 + (A - a)h_3 + b(h_5 - h_3). \tag{10}$$

From (9),

$$h_3 - h_5 = \frac{M}{b\rho_w},$$

or

$$h_5 - h_3 = -\frac{M}{b\rho_w}. \tag{11}$$

Substituting (11) into (10), we obtain

$$(A - a)h_1 + ah_2 = ah_4 + (A - a)h_3 - \frac{M}{\rho_w},$$

or

$$(A - a)h_1 - (A - a)h_3 = ah_4 - ah_2 - \frac{M}{\rho_w},$$

or

$$(A - a)(h_1 - h_3) = a(h_4 - h_2) - \frac{M}{\rho_w},$$

and so

$$h_1 - h_3 = \frac{a}{A - a}(h_4 - h_2) - \frac{M}{(A - a)\rho_w}. \tag{12}$$

From (7),

$$h_1 - h_2 = \frac{W + M}{a\rho_w},$$

and from (8),

$$h_3 - h_4 = \frac{W}{a\rho_w}.$$

These last two results tell us that

$$(h_1 - h_2) - (h_3 - h_4) = \frac{W + M}{a\rho_w} - \frac{W}{a\rho_w} = \frac{M}{a\rho_w},$$

or, with a slight rearrangement on the left-hand side,

$$(h_1 - h_3) - (h_4 - h_2) = \frac{M}{a\rho_w}.$$

So,

$$h_4 - h_2 = \frac{M}{a\rho_w} - (h_1 - h_3).$$

Substituting this equality into (12), we get

$$h_1 - h_3 = \frac{a}{A - a}\left[\frac{M}{a\rho_w} - (h_1 - h_3)\right] - \frac{M}{(A - a)\rho_w}$$

$$= \frac{M}{(A - a)\rho_w} - \frac{a}{A - a}(h_1 - h_3),$$

$$-\frac{M}{(A - a)\rho_w} = -\frac{a}{A - a}(h_1 - h_3),$$

which, because $\frac{a}{A-a} \neq 0$, means that $h_1 - h_3 = 0$, and so $h_1 = h_3$. That is, the lake water level doesn't change if Bob *floats* instead of sinking.

To end this chapter on Archimedes' principle, I'll next show you an interesting "simple physics" question that you can, if you're inclined, experimentally study in your kitchen sink. I first came across it as a challenge problem in the *American Journal of Physics*, but, unfortunately, the creator of the problem solved it incorrectly.[3] Fortunately, a reader published a correct analysis a few months later, and I'll use a variation of his approach here.[4] As you'll quickly appreciate, no "facile word solution" will suffice now, but some solid math (but high school algebra and freshman calculus, only!) is a requirement.

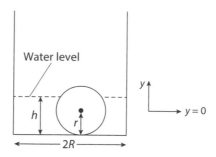

Figure 15.4. A "just-floating" sphere in a cylindrical tank

Imagine that you have an empty cylindrical tank of given radius R, and that you place a sphere of unknown radius r on the bottom of the tank. (Obviously, $r < R$ or else the sphere wouldn't fit.) The sphere has density ρ, the value of which you don't know *except* that it is less than that of water; that is, if you now begin to add water to the tank, the sphere will eventually float. The problem is to compute the amount of water that *just* lifts the sphere off the bottom of the tank. Figure 15.4 shows the geometry of this problem, where as the sphere just starts to float, the water level in the tank is h. In the figure the water level is shown as above the center of the sphere, but the "barely floating" value for h clearly depends on ρ and r. Let's agree as we start the analysis that we choose our units so that the density of water is 1, which means that $0 < \rho < 1$.

If we write v for the volume of water in the tank, and v_s for the volume of the *submerged* portion of the sphere, then

$$v = \pi R^2 h - v_s. \tag{13}$$

There are two possibilities for the depth of the water, h, namely $h \geq r$ (as shown in Figure 15.4) and $h < r$. For $h \geq r$ we can write

$$v_s = \frac{2}{3}\pi r^3 + \int_0^{h-r} \pi (r^2 - y^2) \, dy = \pi \frac{3rh^2 - h^3}{3}, \quad h \geq r,$$

where the first term on the right is the volume of the lower half of the sphere, and the integral[5] is the volume of the portion of the sphere above the sphere's center that is also submerged. The variable y is distance measured from the sphere's center (which is at $y = 0$).

Notice two things about the expression for v_s. First, it gives the correct result for $h = 2r(v_s = \frac{4}{3}\pi r^3)$, and second, it gives the correct answer for the $h < r$ case, too. You should set up the v_s integral directly for the $h < r$ case and verify this (and I *know* you will.) So, to go with (13) we have

$$v_s = \pi \frac{3rh^2 - h^3}{3}, \quad 0 \le h \le 2r. \tag{14}$$

Next, from Archimedes' principle we know that when the sphere just floats it will displace an amount of water equal to the sphere's weight, and so, since the density of water is 1, we have

$$\frac{4}{3}\pi r^3 \rho = \pi \frac{3rh^2 - h^3}{3}, \tag{15}$$

where the left-hand side of (15) is the weight of the sphere, and h on the right-hand side is the water depth when the sphere *just* floats. With some simple algebra we can write (15) as

$$r^3 - r\frac{3h^2}{4\rho} + \frac{h^3}{4\rho} = 0. \tag{16}$$

Okay, the BIG question is, what do we *do* with (16)?

We *could* follow the lead of the author in note 4, who *analytically* solved for the three roots of the cubic equation (16), showing that for $\rho < 1$ there are three *real* solutions of which only one is "physical." (The author doesn't define what he means by *physical*, and I'll tell you more about that in just a bit.) The algebra involved in solving (16) does, however, get fairly complicated (although a very good high school honors-math student could follow it), and so I'll take a different approach.

To start, here's a repeat of a simple observation I made earlier: the largest sphere you can place on the bottom of the tank has radius $r = R$. So, for a given value of $\rho < 1$, let's iteratively solve (on a computer[6]) for r as we let h vary from 0.01 R to 2R, in steps of 0.01R.

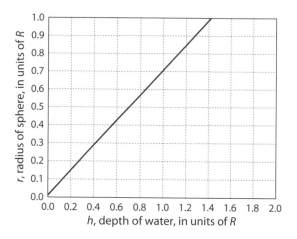

Figure 15.5. r versus h for $\rho = 0.8$

That is, we let h vary from hardly any water in the tank to there being enough water in the tank to just totally submerge the entire sphere, even for the largest possible sphere. Somewhere in that interval for h, any sphere that fits in the tank *will* just start to float. If we pick R to be our unit of length, then h will vary from 0.01 to 2, in steps of 0.01, giving us 200 values of h.

For each of those 200 values of h we'll solve equation (16) for r, and so we'll have 200 pairs of (r, h) values, allowing us to plot a curve of r versus h for the given ρ. Figure 15.5 shows such a plot for the case of $\rho = 0.8$ (chosen arbitrarily), and, as you can see (as an example), if $h = 1$ (in units of R) the radius of the sphere with that density that just begins to float is $r = 0.7$ (more accurately, 0.7014), again in units of R.

Now, here's a little puzzle for you (that we'll quickly answer): there's *another* value of r for $h = 1$ that also satisfies (16), namely, $r = 0.4033$. You can verify this by simply plugging $r = 0.4033$ into (16) along with $h = 1$ and $\rho = 0.8$ and seeing that they work. So why isn't that value of r shown in Figure 15.5? Because it's not "physical!" Here's why.

All cubic equations with real coefficients (as is (16)) have three solutions, each of which is either real or are one of a complex-conjugate *pair*.[7] So, (16) will have either one real solution and two complex solutions, or it will have three real solutions. Having two real and one complex solution is impossible, because complex solutions appear in pairs. Now, it is not difficult to show that a cubic equation of the form

$$r^3 - pr + q = 0,$$

where both p and q are positive (as in (16)), *always* has a *negative* real solution,[8] a solution that we'd immediately reject as *unphysical*. (After all, when is the last time you saw a sphere with a negative radius?) That means, from the first sentence of this paragraph, that the other two solutions are either both complex or both real.

If those two solutions are complex then we'd reject them, too, as unphysical, because a complex radius (shades of the fourth dimension!) is at least as bad as a negative radius. But this possibility will not occur for our (16) because, *physically*, we know that for every value of h there *must* be some sphere (some r) that *will* float. So, we know that (16) *will* have three real solutions. Further, the analysis mentioned in note 8, showing that there is always one negative solution, also shows that the other two real solutions must *both* be positive.

The fact that those two real solutions are positive is, however, *not* sufficient to allow them to pass the "physical" requirement. There are, in fact, *two* additional requirements a positive solution has to satisfy to be declared physical. First, the positive r must be no larger than 1 (in units of R) or else, as mentioned earlier, the sphere won't fit in the tank. Well, you say, *both $r = 0.7014$ and $r = 0.4033$* are less than 1, so *both* pass that test. But there is one final test of physicality that $r = 0.4033$ does *not* pass. Have you seen it yet?

To be a physically valid solution, a positive value of r satisfying (16) must be such that $h < 2r$. That simply means the sphere starts floating *before* it is totally submerged. If it hasn't floated by the time it's totally submerged, it isn't suddenly going to float just because you add even more water to the tank! The $r = 0.4033$ solution flunks that test because $h = 1 > 0.8066$. The $r = 0.7014$ solution shown in Figure 15.5 *does* pass this final test of physicality ($h = 1 < 1.4028$). The bottom line is that for every h and $\rho < 1$ there is exactly *one* physically valid value of r.

Now, what about our original question: what's the amount of water in the tank when the sphere just begins to float? We can answer that once we have r for a given ρ and h. We just substitute r and h into (14) to get v_s, and then substitute v_s into (13) to get v. That's it!

Notes

1. A very nice discussion of how physicists should discuss what Archimedes did (whatever it was) is in Lillian Hartmann Hoddeson, "How Did Archimedes Solve King Hiero's Crown Problem?—An Unanswered Question," *The Physics Teacher*, January 1972, pp. 14–18.

2. Ironically, a reading of Vitruvius strongly hints at buoyancy having *nothing* to do with the king's crown problem! Vitruvius writes: "[Archimedes] chanced to come to the place of bathing, and there, as he was sitting down in the tub, he noticed that the amount of water which flowed over the tub was equal to the amount by which his body was immersed." So, according to Vitruvius, what Archimedes discovered is that a way to measure the volume of a complicated object (like the king's crown) is by having it displace water. (Does this remind you of the Edison story I told you at the end of Chapter 1?) For a given weight, gold and silver will displace different volumes because their densities are different. With this approach, it is displaced volume, not buoyancy that is the key idea.

3. I. Richard Lapidus, "Floating Sphere," *American Journal of Physics*, March 1985, pp. 269 and 280.

4. Lawrence Ruby, "Floating Sphere Problem," *American Journal of Physics*, November 1985, pp. 1035–1036.

5. I won't go through the details of this integral other than to say it is a standard volume integral example in virtually every freshman calculus textbook. You should go through the details of *evaluating* the integral, however, just to verify the result.

6. This is a book on physics, not computer programming, but if you're curious, I used the symbolic math features of MATLAB. If you're *really* curious about the details, write to me and I'll send you the code that generated Figure 15.5.

7. This is the conclusion from a purely *mathematical* argument from the theory of equations (no physics involved), and you can find more on it in books on that subject. As physicists we trust in our mathematician friends and take the conclusion as fact.

8. Can you prove this? It's pure math (no physics), and if you can't (but you're curious), write to me and I'll send you an analysis that is neither long nor difficult.

16. A Reciprocating Problem

The wheels on the bus go round and round . . .
The people on the bus go up and down . . .
— *lyrics from a nursery-school song that has driven the parents of
prekindergarten kids nuts for decades*

For an example of the use of trigonometry, geometry (and some calculus, too) to attack an important engineering-physics problem, consider Figure 16.1. There you see a cross-sectional view of a rotating *crankshaft* at A, with a crank arm of length r extending out to a hinged joint at B. As the crankshaft rotates counterclockwise at the constant angular speed of ω radians/second, B rotates along the circumference of a circle with radius r at a constant speed. B, in turn, is linked to a hinged joint at C via a *connecting rod* of length l; C is the location of a *wrist pin* that allows an attached piston to be driven back and forth along the x-axis by the connecting rod.

As described, the piston moves *because* the crankshaft is turned by some external energy source (say, a turbine submerged in running water) and so the entire arrangement could be a pump. On the other hand, the crankshaft could be rotating (and so driving the transmission and, hence, the wheels of a car) *because* the piston is powered by rapidly burning gasoline vapor in a cylinder that encloses the piston. In this case we have an internal combustion engine. In any case, given that the crankshaft is rotating we are to calculate the position, speed, and acceleration of the piston's wrist pin.

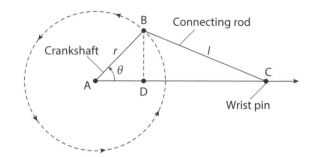

Figure 16.1. Crankshaft/connecting rod/wrist pin geometry

From the geometry shown in Figure 16.1 we can write the wrist pin location, measured from A, as

$$x(t) = \overline{AD} + \overline{DC}.$$

Notice, carefully, that we write $x = x(t)$ because $\theta = \theta(t) = \omega t$. Now, since

$$\overline{AD} = r \cos(\theta),$$

and since the Pythagorean theorem tells us that

$$\overline{BD}^2 + \overline{DC}^2 = l^2,$$

where

$$\overline{BD} = r \sin(\theta),$$

we have

$$x(t) = r \cos(\theta) + \sqrt{l^2 - r^2 \sin^2(\theta)} = r \cos(\theta) + l\sqrt{1 - \left(\frac{r}{l}\right)^2 \sin^2(\theta)},$$

or

$$\boxed{\frac{x(t)}{l} = \left(\frac{r}{l}\right) \cos(\theta) + \sqrt{1 - \left(\frac{r}{l}\right)^2 \sin^2(\theta)}, \quad \theta = \omega t.}$$

The boxed equation for $\frac{x(t)}{l}$ shows the useful technique of *normalizing* variables: we have the position of the wrist pin relative to the length of the connecting rod; that is, the length of the connecting rod is playing the role of the unit length.

To find the speed of the wrist pin, we differentiate the expression for $x(t)$—*not* the normalized $\frac{x(t)}{l}$—to get

$$\frac{dx}{dt} = -r\sin(\theta)\frac{d\theta}{dt} + \frac{1}{2}\left\{l^2 - r^2\sin^2(\theta)\right\}^{-1/2}\left\{-2r^2\sin(\theta)\cos(\theta)\frac{d\theta}{dt}\right\},$$

which, after a bit of simple algebra, becomes

$$\frac{dx}{dt} = -\omega r\sin(\theta) - \frac{\omega r r\sin(\theta)\cos(\theta)}{l\sqrt{1 - \left(\frac{r}{l}\right)^2\sin^2(\theta)}}.$$

The speed of B is such that in one complete revolution B moves through distance $2\pi r$ in $\frac{2\pi}{\omega}$ seconds, and so B's speed is

$$\frac{2\pi r}{\frac{2\pi}{\omega}} = \omega r,$$

which we'll use as the unit of speed to normalize the speed of the wrist pin. That is, the normalized wrist pin speed is

$$\frac{\frac{dx}{dt}}{\omega r} = -\sin(\theta)\left\{1 + \frac{\left(\frac{r}{l}\right)\sin(\theta)\cos(\theta)}{\sqrt{1 - \left(\frac{r}{l}\right)^2\sin^2(\theta)}}\right\}, \qquad \theta = \omega t.$$

Finally, to get the acceleration of the wrist pin, we'll differentiate $\frac{dx}{dt}$ to get

$$\frac{d^2x}{dt^2} = -\omega r\cos(\theta)\frac{d\theta}{dt} - r^2\omega$$

$$\times \left[\frac{\sqrt{l^2 - r^2\sin^2(\theta)}\left\{\cos^2(\theta)\frac{d\theta}{dt} - \sin^2(\theta)\frac{d\theta}{dt}\right\} - \sin(\theta)\cos(\theta)\frac{1}{2}}{\times \left\{l^2 - r^2\sin^2(\theta)\right\}^{-1/2}\left\{-2r^2\sin(\theta)\cos(\theta)\frac{d\theta}{dt}\right\}}{l^2 - r^2\sin^2(\theta)}\right],$$

which, after a bit of simple algebra, reduces to

$$\frac{d^2x}{dt^2} = -\omega^2 r\left[\cos(\theta) + \left(\frac{r}{l}\right)\frac{\cos(2\theta) + \left(\frac{r}{l}\right)^2\sin^4(\theta)}{\left\{1 - \left(\frac{r}{l}\right)^2\sin^2(\theta)\right\}^{3/2}}\right].$$

As we've done twice before, we normalize this acceleration with an acceleration inherent in the problem, and here that is $\omega^2 r$ (which you can check has the units of acceleration;[1] earlier in the book, in Chapter 5, we called this the *centripetal* acceleration). So, the normalized acceleration of the wrist pin is

$$\frac{\frac{d^2x}{dt^2}}{\omega^2 r} = -\left[\cos(\theta) + \left(\frac{r}{l}\right)\frac{\cos(2\theta) + \left(\frac{r}{l}\right)^2\sin^4(\theta)}{\left\{1 - \left(\frac{r}{l}\right)^2\sin^2(\theta)\right\}^{3/2}}\right], \quad \theta = \omega t.$$

Figure 16.2 shows plots of our three boxed expressions for the wrist pin's normalized position, speed, and acceleration, for two values of the normalized parameter $\frac{r}{l}$ ($\frac{1}{2}$ in the left column, and $\frac{1}{3}$ in the right column). The independent variable, the angle θ, is plotted on the horizontal axes for one complete revolution of the crankshaft, rather than time, as that is the parameter automakers use to specify the proper setting for the ignition timing in their internal combustion engines. For example, in specification sheets for timing, mechanics will find phrases like "set at 12 degrees BTDC," which translates as "set the spark plug to fire when the piston is in the position 12 degrees before top dead center of the compression stroke."

These plots would be of great interest to the mechanical design engineers responsible for selecting the metals with the necessary strength to withstand the expected speeds and accelerations of the crankshaft/connecting rod/wrist pin assembly.

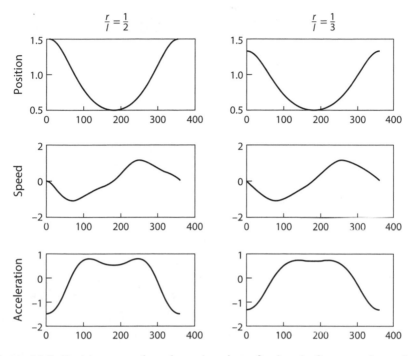

Figure 16.2. Position, speed, and acceleration of wrist pin for two values of r/l

Note

1. The units of $\omega^2 r$ are $\frac{\text{radians-squared·meters}}{\text{seconds squared}}$, but *radians* are considered to be physically dimensionless.

17. How to Catch a Baseball (or Not)

Physical Laws should have mathematical beauty.
— *written on a Moscow blackboard in 1955 by the 1933 Nobel*
Prize in Physics winner Paul Dirac

In this discussion you'll see an amazing theoretical result from trigonometry and physics that seems to explain an equally amazing (but surprisingly routine) occurrence in baseball. Alas, the "explanation" is false. This doesn't deny Dirac's thesis (good physics should be beautiful) but, rather, makes the point that the converse (beautiful physics is good physics) is not necessarily the case.

And that's too bad, because the theory I'll explain next *is* remarkably beautiful in its simplicity. The problem has its origin in an essay published by the American electrical engineer Vannevar Bush (1890–1974), "When Bat Meets Ball." There he wrote "Willie Mays, at the crack of the bat, will take a brief look at the flight of the ball, run without looking back, be at exactly the right spot at the right time, and take the ball over his shoulder with a basket catch. How he does it no one knows, certainly not Willie Mays."[1]

Even for those who (like me) find baseball to be a game in which every game looks a lot like the last one, this one particular athletic feat *is* something to see. The very next year after Bush's essay appeared, however, one analyst thought he had reduced it to pure mathematics, and he wrote "It does not seem entirely mysterious."[2] Declaring the problem to be simply one of "predicting the motion of a target when its laws of motion are known," and that such a prediction "is a standard one for astronomers [and] ballistic-missile defense engineers," Chapman argued as follows.

"Let the ball leave the bat (the origin) with an initial speed of V at an angle θ with the ground. As is well known … the vertical and horizontal displacements [that is, the x- and y-coordinates of the ball] at any time t [$t = 0$ is the instant the batter hits the ball] are

$$ y = V\sin(\theta)t - \frac{1}{2}gt^2, $$

$$ x = V\cos(\theta)t, $$

where g is the magnitude of the acceleration of gravity."

Notice, carefully, that *if we ignore air resistance* (as did Chapman), the only force acting on the ball once the ball leaves the bat is gravity, *vertically downward*, and so the *horizontal* component of the ball's speed—$V\cos(\theta)$—never changes. Thus, we get the preceding equation for x. For the *vertical* component of the ball's speed, however, gravity makes itself felt by continually reducing that initial speed component—$V\sin(\theta)$—and so we have

$$ \frac{dy}{dt} = V\sin(\theta) - gt, $$

which easily integrates to Chapman's equation for y.[3]

Chapman next asked his readers to consider Figure 17.1. The batter is at the origin of the x, y–coordinate system, and the fielder is (luckily) standing right where the ball will eventually land (this special condition will be somewhat relaxed in just a bit), distance R from the batter. Thus, the fielder does not actually see the arc of the ball's trajectory, but, instead, the ball appears to him to be simply first rising and then falling in a vertical plane that passes through the fielder and the batter. What visual cue to the fielder can there be in this situation— the toughest one that a fielder can face—that tells him that the ball is coming right to him? This is the question that Chapman thought he answered.

In Figure 17.1 the fielder's line of sight to the ball makes angle ϕ with respect to the ground, and the fielder is distance R from the batter (where R is point at which the ball will return to Earth). Chapman doesn't give any intermediate details (writing only that after "modest algebraic manipulation" of the x- and y-equations his answer results), but I'll show you next what he did.

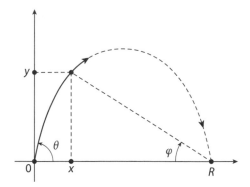

Figure 17.1. A hit ball going right to the fielder

To start, we define $t = T$ to be the time when the ball returns to Earth (that is, when the fielder catches the ball). Then, as $y(T) = 0$, we have

$$V \sin(\theta) T - \frac{1}{2} g T^2 = 0,$$

and solving for $T > 0$, we get

$$T = \frac{2 V \sin(\theta)}{g}.$$

Substituting this result into the equation for x, and since $x(T) = R$, we have

$$R = \frac{2 V^2 \sin(\theta) \cos(\theta)}{g}.$$

From the geometry of Figure 17.1 we can immediately write, for every instant of time $0 < t < T$,

$$\tan(\phi) = \frac{y}{R - x} = \frac{V \sin(\theta) t - \frac{1}{2} g t^2}{\frac{2 V^2 \sin(\theta) \cos(\theta)}{g} - V \cos(\theta) t} = \frac{t \left[V \sin(\theta) - \frac{1}{2} g t \right]}{V \cos(\theta) \left[\frac{2 V \sin(\theta)}{g} - t \right]}$$

$$= \frac{t [2 V \sin(\theta) - g t] \frac{1}{2}}{V \cos(\theta) \frac{1}{g} [2 V \sin(\theta) - g t]},$$

$$= \frac{g}{2 V \cos(\theta)} t,$$

and so we arrive at the simple result

$$\tan(\phi) = (\text{constant})t.$$

That is, for a fielder standing right where the ball will land, the tangent of his line-of-sight elevation angle to the ball's instantaneous location increases linearly with time.

Now, before examining what this admittedly pretty mathematical result might actually mean, let's next see what happens in the more realistic case of the fielder *not* fortuitously standing in the right spot for a motionless catch. Suppose instead that the fielder still sees the ball in the vertical plane containing himself and the batter, but now he is distance s from where the ball will land. That is, at time $t = 0$ the fielder is either distance $R - s$ from the batter or distance $R + s$ from the batter. I'll do the "too close" case here—and so the fielder will have to run outward (away) from the origin—and I'll let you make the very minor alterations in the analysis to show that the "too far" case leads to the same result.

Suppose that τ is the fielder's reaction time and that once he decides he has to move, the fielder runs at the constant speed v that just gets him to $x = R$ at time $t = T$, that is,

$$s = v(T - \tau).$$

The fielder's coordinate along the horizontal axis at time $t \geq \tau$ is $(R - s) + v(t - \tau)$, and so now we can write

$$\tan(\phi) = \frac{y}{(R - s) + v(t - \tau) - x}.$$

Since

$$s = vT - v\tau,$$

then

$$\tau = \frac{vT - s}{v} = T - \frac{s}{v},$$

and so

$$\tan(\phi) = \frac{V\sin(\theta)t - \frac{1}{2}gt^2}{R - s + v\left(t - T + \frac{s}{v}\right) - V\cos(\theta)t}$$

$$= \frac{t\left[V\sin(\theta) - \frac{1}{2}gt\right]}{\frac{2V^2\sin(\theta)\cos(\theta)}{g} - s + v(t - T) + s - V\cos(\theta)t}$$

$$= \frac{t[2V\sin(\theta) - gt]\frac{1}{2}}{\frac{2V^2\sin(\theta)\cos(\theta)}{g} + v\left[t - \frac{2V\sin(\theta)}{g}\right] - V\cos(\theta)t}$$

$$= \frac{\frac{1}{2}gt[2V\sin(\theta) - gt]}{2V^2\sin(\theta)\cos(\theta) + v[gt - 2V\sin(\theta)] - Vg\cos(\theta)t}$$

$$= \frac{\frac{1}{2}gt\,[2V\sin(\theta) - gt]}{2V^2\sin(\theta)\cos(\theta) - v[2V\sin(\theta) - gt] - Vg\cos(\theta)t}$$

$$= \frac{\frac{1}{2}gt[2V\sin(\theta) - gt]}{V\cos(\theta)[2V\sin(\theta) - gt] - v[2V\sin(\theta) - gt]}$$

$$= \frac{\frac{1}{2}gt[2V\sin(\theta) - gt]}{[2V\sin(\theta) - gt][V\cos(\theta) - v]}$$

$$= \frac{gt}{2[V\cos(\theta) - v]},$$

or, once again,

$$\tan(\phi) = (\text{constant})t.$$

So, just as before, even with the added complications of the two new variables s and τ, the tangent of the fielder's line-of-sight elevation angle to the instantaneous location of the ball increases linearly with time. Amazing!

But, you might wonder after a little thought, so what? Chapman writes at the end of his analysis that "Obviously no ball player ever solves trigonometric equations to catch a ball. What I have tried to show here is that the astonishingly simple amount of information on the constancy of the rate of change of $\tan(\phi)$ . . . tells him that he is running at the right speed in the right direction for the catch." But how does that explain Willie Mays running *with his back to the ball and so not looking at it* until just before the catch? And there is another serious objection to Chapman's analysis. He ignored air resistance, writing (incorrectly), "Aerodynamic forces on a baseball are relatively small and have only a small percentage effect on the trajectory." That is simply not true, and his x- and y-equations are incomplete right from the start. They both require an additional air drag term, and their absence completely invalidates his admittedly beautiful $\tan(\phi)$ result. Pretty, yes, but (unless you're playing on the Moon in a vacuum) wrong.[4]

Notes

1. In Bush's book *Science Is Not Enough*, William Morrow & Company, 1967, pp. 102–122. Baseball Hall of Fame member Willie Mays was, of course, the great center fielder for the New York and San Francisco Giants (and then the New York Mets) from 1951 to 1973.

2. Seville Chapman, "Catching a Baseball," *American Journal of Physics*, October 1968, pp. 868–870.

3. As I typed this, I was reminded of a story that University of California–Santa Barbara physics professor Anthony Zee tells in his book *Einstein Gravity in a Nutshell*, Princeton University Press, 2013, p. 501, when reminiscing about his undergraduate days at Princeton: "When I was a freshman, it was announced that [the eminent Princeton physics professor] John Wheeler would give an experimental (in the sense of pedagogy rather than physics) course to a handpicked group of beginning students. Wheeler asked the group of assembled students a series of questions to separate the goats from the elect, so to speak. I still remember the question that eliminated the largest number of hopefuls. Does a tossed ball have zero acceleration at the top of its flight?"

The answer (which I am guessing Zee got right) is, of course, no, the tossed ball is *always* accelerating downward at precisely 1 gee. In Chapman's problem, the same is true, as $\frac{d^2y}{dt^2} = -g$.

4. For how to properly handle air drag in Chapman's analysis (it isn't trivial!), see Peter J. Brancazio, "Looking into Chapman's Homer: The Physics of Judging a Fly Ball," *American Journal of Physics*, September 1985, pp. 849–855. This paper also discusses, in some detail, what the visual cues to a fielder might actually be.

18. Tossing Balls and Shooting Bullets Uphill

Does the road wind up-hill all the way?
Yes, to the very end.
Will the day's journey take the whole long day?
From morn to night, my friend.
— *the problems of going "uphill" transcend physics*[1]

A high school physics teacher in Pennsylvania observes a physical education class throwing a softball along a surface that slopes upward, as part of a national program to determine physical fitness.[2] When told that a student's performance is the distance the ball goes until it hits the ground, the teacher quickly realizes that the students are being evaluated incorrectly. He goes home to think some more about it.

Years later, a high school physics teacher in Norway is asked in class by a student, "Is it true that you always hit too high when shooting [a rifle] uphill [at a deer]?"[3] The resulting class discussion quickly leads the students to conclude that yes, you shoot high uphill (and low downhill), but the teacher isn't so sure that's completely correct. She goes home to think some more about it.

These two seemingly quite different situations involve the same simple physics, and we can use Figure 18.1 to simultaneously model the geometry of both the problems presented to our high school teachers. To gain an understanding of what is going on in the two problems, all we'll need in the way of math is some pretty straightforward trigonometry and just the briefest touch of freshman calculus. For both

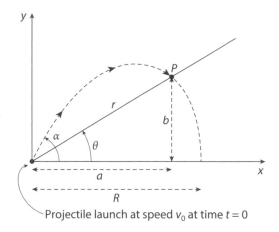

Figure 18.1. The geometry of tossing/shooting uphill

problems, we'll ignore the complications of air drag and will consider gravity as the only force acting on either the softball or the bullet as each travels its path.

For the softball problem, a student's performance is recorded as the value of r, when it should really be the value of R. That is, when the national standards for the softball toss were established they were for a toss over a horizontal surface ($\theta = 0$), not for one inclined at an upward angle ($\theta > 0$). The Pennsylvania teacher's problem was, given the measured r and the value of θ, to determine a correction formula that gives R as a function of r, θ, and α (the angle of the initial toss *measured with respect to the horizontal*).

For the shooting problem, the hunter has almost certainly sighted his or her rifle (to compensate for bullet drop) over the particular distance at which he or she expects to engage a target—and has practiced firing it—at a gun range that is horizontal. The Norwegian teacher's problem is to determine what effect $\theta \neq 0$ has ($\theta > 0$ models shooting uphill, and $\theta < 0$ models shooting downhill) on the location of the bullet's impact point, P.

To start our analysis, let's write the initial speed of the *projectile* (either the softball or the bullet) as v_0. In the x, y-coordinate system shown in Figure 18.1, we can write the initial ($t = 0$) speed components of the projectile as

$$v_{0x} = v_0 \cos(\alpha),$$

and

$$v_{0y} = v_0 \sin(\alpha).$$

Since gravity only acts downward on the projectile, only the y-component of the speed is affected; the x-component is unchanged. So, using g for the acceleration of gravity, we can write the speed components of the projectile at time $t \geq 0$ as

$$v_x(t) = v_{0x} = v_0 \cos(\alpha) = \frac{dx}{dt},$$

and

$$v_y(t) = v_{0y} - gt = v_0 \sin(\alpha) - gt = \frac{dy}{dt}.$$

Integrating these last two equations with respect to time, we get the coordinates of the projectile's location at time t:

$$x(t) = v_0 t \cos(\alpha),$$

and

$$y(t) = v_0 t \sin(\alpha) - \frac{1}{2} g t^2,$$

where, of course, I've selected the arbitrary constants of integration to be such that $x(0) = y(0) = 0$, since the starting point for the projectile (in both problems) is the origin.

Solving the $x(t)$ equation for t we get

$$t = \frac{x}{v_0 \cos(\alpha)},$$

and then, substituting this result for t into the y-equation, we obtain the *parabolic* path of the projectile (a discovery due to Galileo, in 1638):

$$y = x \tan(\alpha) - \frac{g}{2 v_0^2 \cos^2(\alpha)} x^2.$$

The x, y-coordinates of P, the impact point of the projectile on the inclined surface, are $x = a$, $y = b$, where

$$a = r\cos(\theta), \quad b = r\sin(\theta).$$

Substituting these expressions for x and y into the parabolic path equation, we get

$$r\sin(\theta) = r\cos(\theta)\tan(\alpha) - \frac{g}{2v_0^2\cos^2(\alpha)}r^2\cos^2(\theta).$$

An obvious, trivial solution for r is $r = 0$, which we promptly ignore. We get a much more interesting result if we factor out an r and rearrange the terms to write

$$r\left[\frac{g}{2v_0^2\cos^2(\alpha)}r\cos^2(\theta) + \sin(\theta) - \cos(\theta)\tan(\alpha)\right] = 0$$

and set the factor in the square brackets to zero. Then,

$$r = \frac{\{\cos(\theta)\tan(\alpha) - \sin(\theta)\}\,2v_0^2\cos^2(\alpha)}{g\cos^2(\theta)}$$

$$= \frac{\cos(\theta)\left\{\tan(\alpha) - \frac{\sin(\theta)}{\cos(\theta)}\right\}2v_0^2\cos^2(\alpha)}{g\cos^2(\theta)}$$

$$= \frac{\left\{\frac{\sin(\alpha)}{\cos(\alpha)} - \frac{\sin(\theta)}{\cos(\theta)}\right\}2v_0^2\cos^2(\alpha)}{g\cos(\theta)} = \frac{\left\{\sin(\alpha) - \frac{\sin(\theta)}{\cos(\theta)}\cos(\alpha)\right\}2v_0^2\cos(\alpha)}{g\cos(\theta)}$$

$$= \frac{\{\cos(\theta)\sin(\alpha) - \sin(\theta)\cos(\alpha)\}2v_0^2\cos(\alpha)}{g\cos^2(\theta)}$$

or, at last, once we remember the trigonometric identity for the difference of two angles,

$$r = \frac{2v_0^2}{g\cos^2(\theta)} \cos(\alpha)\sin(\alpha - \theta). \tag{A}$$

The expression in (A) gives the distance of the softball toss measured along the inclined surface. If $\theta = 0$, which means the toss is over a horizontal surface, then $r = R$, and so

$$R = \frac{2v_0^2}{g} \cos(\alpha)\sin(\alpha). \tag{B}$$

From (A) we have

$$\frac{2v_0^2}{g} = \frac{r\cos^2(\theta)}{\cos(\alpha)\sin(\alpha - \theta)}$$

and substituting this expression into (B) gives us our Pennsylvania high school physics teacher's conversion equation:

$$R = r\frac{\cos^2(\theta)\sin(\alpha)}{\sin(\alpha - \theta)}. \tag{C}$$

To use (C), we of course have to first decide what tossing angle α to use. The best choice is that value of α that maximizes r, which we can find by setting the derivative of r with respect to α to zero. Doing that, we get, from (A),

$$\frac{dr}{d\alpha} = \frac{2v_0^2}{g\cos^2(\theta)}[\cos(\alpha)\cos(\alpha-\theta) - \sin(\alpha)\sin(\alpha-\theta)]$$

$$= \frac{2v_0^2}{g\cos^2(\theta)}\cos(2\alpha-\theta) = 0.$$

Thus,

$$2\alpha - \theta = 90°,$$

or the value of α that gives the maximum distance up the inclined surface at angle θ is

$$\alpha = 45° + \frac{1}{2}\theta.$$

For a horizontal surface ($\theta = 0°$) we find $\alpha = 45°$ is best, but for an incline of $2°$ (for example), the slightly larger $\alpha = 46°$ is best. So, suppose a student tosses a softball up a $2°$ incline a distance of $r = 200$ feet. What *should* be recorded, for purposes of national comparison, is

$$R = 200\frac{\cos^2(2°)\sin(46°)}{\sin(44°)}\text{feet} = 207\text{ feet},$$

a not insignificant correction.

Okay, let's now turn to the problem posed to the Norwegian physics teacher. Returning to (B), we see that if the rifle is sighted at a *horizontal* gun range to precisely hit a target at distance R, then the rifle has to be elevated above the horizontal shooting surface by the angle ϕ, where

$$\cos(\phi)\sin(\phi) = \frac{Rg}{2v_0^2} = \frac{1}{2}\sin(2\phi),$$

and so

$$\sin(2\phi) = \frac{Rg}{v_0^2}.$$

Referring to Figure 18.1, we see that ϕ is α for the special case of $\theta = 0$. The angle ϕ is, generally, not large. For example, a .30-06 ("thirty aught-six") bolt-action hunting rifle might have a muzzle velocity of around 2,500 feet/second, and so, for a target at $R = 200$ yards (600 feet), the elevation angle ϕ would be

$$\phi = \frac{1}{2} \sin^{-1}\left\{ \frac{Rg}{v_0^2} \right\} = \frac{1}{2} \sin^{-1}\left\{ \frac{600 \times 32.2}{2,500^2} \right\} = 0.089°.$$

Now, for shooting uphill, let's suppose the rifle is elevated by angle β, so that the value of r is still R, the distance at which the rifle was sighted (with angle ϕ) on a horizontal shooting surface. How does β compare with ϕ? We have $\alpha = \theta + \beta$, and so, from (A),

$$R = \frac{2v_0^2}{g\,\cos^2(\theta)}\,\cos(\theta + \beta)\sin(\beta),$$

or, since

$$R = \frac{2v_0^2}{g}\,\sin(2\phi),$$

we have

$$\sin(2\phi) = \frac{2\cos(\theta + \beta)\sin(\beta)}{\cos^2(\theta)}.$$

That is,

$$\frac{1}{2}\sin(2\phi)\cos(\theta) = \frac{\{\cos(\theta)\cos(\beta) - \sin(\theta)\sin(\beta)\}\sin(\beta)}{\cos(\theta)}$$

$$= \cos(\beta)\sin(\beta) - \tan(\theta)\sin^2(\beta)$$

$$= \frac{1}{2}\sin(2\beta) - \tan(\theta)\sin^2(\beta),$$

and so, finally,

$$\boxed{\sin(2\beta) = \sin(2\phi)\cos(\theta) + 2\tan(\theta)\sin^2(\beta)} \qquad \text{(D)}$$

There is a lot of information tucked away in (D). Notice, first, that if $\theta = 0$ (a horizontal gun range), then $\tan(\theta) = 0$, and $\cos(\theta) = 1$, and so $\beta = \phi$, just as it should be. Also notice that if $\theta \neq 0$, then, because $\cos(\theta)$ is even about $\theta = 0$, while $\tan(\theta)$ is odd about $\theta = 0$, $\sin(2\beta)$ will be equal to $\sin(2\phi)\cos(\theta)$ *plus* a correction term if $\theta > 0$ (shooting uphill) but minus that same correction term if $\theta < 0$ (shooting downhill). That is, for the shooter to hit the target at the same distance R from him- or herself on an incline, the elevation angle β is different for the uphill and downhill scenarios.

But, since elevation angles for a high-velocity rifle are small, the correction term is also small (if β is small, then $\sin(\beta)$ is small, and $\sin^2(\beta)$ is *very* small). So, let's ignore that quite small correction term and simply write

$$\sin(2\beta) = \sin(2\phi)\cos(\theta),$$

which immediately tells us that $\beta < \phi$. For example, for the .30-06 rifle with a muzzle velocity of 2,500 feet/second considered earlier, for shooting at a target 600 feet away on a 35° incline, the elevation angle should be

$$\beta = \frac{1}{2}\sin^{-1}\left\{\frac{Rg}{v_0^2}\cos(35°)\right\} = \frac{1}{2}\sin^{-1}\left\{\frac{600 \times 32.2}{2,500^2} \times 0.81915\right\} = 0.0725°.$$

This reduction in the elevation angle is necessary because a person using the elevation angle ϕ calculated earlier for a horizontal shot will overshoot the target with an inclined shot. Will he or she overshoot by much? Yes.

To see this, let's approach the problem in a different way. Our last calculation was to find the elevation angle β for hitting the target at the same distance on an incline as at an elevation angle of ϕ does for a horizontal shot. Let's now continue to use the elevation angle ϕ on the incline shot and calculate the distance r at which the bullet hits the incline. We have $\alpha = \theta + \phi$, and so, from (A),

$$r = \frac{2v_0^2}{g\cos^2(\theta)}\cos(\theta + \phi)\sin(\phi),$$

or, since

$$\frac{v_0^2}{g} = \frac{R}{2\cos(\phi)\sin(\phi)},$$

we have

$$r = \frac{\cos(\theta + \phi)}{\cos(\phi)\cos^2(\theta)}R.$$

For $\theta = 35°$ and $\phi = 0.089°$, we have

$$r = \frac{\cos(35.089°)}{\cos(0.089°)\cos^2(35°)}R = \frac{0.81826}{(1)(0.671)}R = 1.22R.$$

You'll recall that $R = 600$ feet, and so $r = 732$ feet, a fairly big overshoot. This will be the case for both the uphill and the downhill scenarios, and so the physics teacher's class was right in its conclusion about the uphill case but wrong in its conclusion about the downhill case.

I'll end this chapter with a couple of historical notes. The problem of shooting a gun on an inclined surface is a very old one, originating not in high school physics classes in Norway or Connecticut but in the early 1640s with the Italian mathematician Evangelista Torricelli (1608–1647). Our result that the value of α that gives the maximum distance up an inclined surface at angle θ is

$$\alpha = 45° + \frac{1}{2}\theta$$

was found a half-century later by Newton's friend Edmund Halley (see note 2 in Chapter 14), who published it in 1695 in the *Philosophical*

Transactions of the Royal Society. There he expressed this result in the following elegant way. Noticing that

$$\alpha = 45° + \frac{1}{2}\theta = \theta + \frac{1}{2}(90° - \theta),$$

Halley observed that the maximum range for shooting on an incline is achieved by shooting at the angle bisecting the angle formed by the inclined surface and the vertical.

Notes

1. These words are the opening stanza to the 1861 poem "Up-Hill" by Christina Rossetti, a major English poet of the Victorian period.

2. Joseph C. Baiera, "Physics of the Softball Throw," *The Physics Teacher*, September 1976, pp. 367–369.

3. Ole Anton Haugland, "A Puzzle in Elementary Ballistics," *The Physics Teacher*, April 1983, pp. 246–248.

19. Rapid Travel in a Great Circle Transit Tube

For my part, I travel not to go anywhere, but to go. I travel
for travel's sake. The great affair is to move.
— Robert Louis Stevenson's 1878 Travels with a Donkey. *In this*
analysis we'll study a means of ground travel in which you would
move a lot *faster than a donkey.*

In the late 1890s and thereafter, one of the more fantastic of that era's "scientifiction" (as science fiction was then called) themes was rapid travel from one city on a planet (not necessarily the Earth) to another, along straight-line tunnels drilled right through the planet. (The 1864 novel *A Journey to the Center of the Earth*, by Jules Verne, in which characters simply *climb down* to their destination, is more a romantic fantasy than it is a work of science fiction.) The most extreme form of this theme, the schoolboy dream of a "hole to China," that is, a tunnel along a diameter of the planet through the core and out the other side, appeared in a 1929 novel called *The Earth-Tube*. Much more realistic would be a straight tunnel joining, say, New York City and Philadelphia, a tunnel that would, at its deepest, dip to "just" 1,200 feet below the surface.[1]

In 1953 a far more realistic transportation system concept was proposed,[2] one that I don't believe attracted much attention. I find it quite interesting 62 years later, though (as I write), and perhaps it was simply ahead of its time. The mathematical analysis of this concept will be the most challenging chapter in the book, but if you stay the course I think you'll find it well worth the effort.

If someone asks you, what is the shortest path connecting two points on a flat surface? I'm sure you'd quickly answer, a straight line. But what would you say to the same question if the surface were not flat but instead spherical (like the Earth's surface)? The straight-line path is a tunnel *through* the Earth, but, as I hinted at in the opening paragraphs, we aren't going to go that route (yes, yes, a bad pun, I know). The answer is a *great circle* path, that is, the intersection of the curved surface with a plane that passes through the two given points *and the center of the sphere*. There are, of course, *two* great circle paths connecting two given points on a sphere (going in opposite directions), and we are talking here about the shorter of the two possibilities, what mathematicians call the *minor arc*.[3] Great circle routes are of great interest to airlines in their attempts to minimize the flight times and fuel costs of trips.

Suppose, to be specific, we want to travel from New York City to Melbourne, Australia, a journey nearly halfway around the planet on a great circle path of length 10,000 miles or so. To make this trip by commercial jet is a long, exhausting task, requiring 20 hours in the air. What if, instead, you could do it in just 44 minutes and yet never be very far from the surface (no rocket journey involved)? That would really be something, don't you think? Some simple physics (and some math) will show us that it's not at all impossible to do from a scientific standpoint, although it will probably not be cheap.

Imagine an elevated transit tube, one evacuated of air, through which a passenger vehicle can move along a great circle path from start to finish. By *elevated* I mean a tube supported by towers that rides several tens of feet above the Earth's surface. The tube encloses a vacuum because the passenger car will be moving, as we'll find, at speeds of several miles/*second*. The value of a great circle path, besides its economic virtue of being the shortest possible surface path, is the *mathematical* fact that, at all times, the gravitational and centrifugal forces on the vehicle and its passengers are radial in direction (but in opposite directions, naturally). There is, of course, a third force also acting on the vehicle, the one that propels the vehicle through the transit tube and produces the acceleration $\frac{d^2s}{dt^2}$ *tangent* to the Earth's surface. Finally, we'll ignore any effects due to a *rotating* Earth.

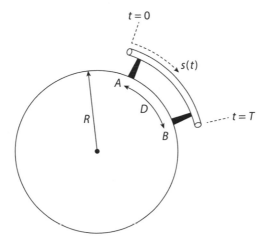

Figure 19.1. Geometry of the transit tube on a nonrotating Earth (not to scale!)

Figure 19.1 represents the Earth, along with a transit tube connecting points A and B. Note, carefully, that because of the symmetry of a sphere we can always position the sphere so that, without loss of generality, the transit-tube geometry looks as shown in the figure. If $s(t)$ is the distance at time t that the vehicle has traveled from A on its way to B, then $s(0) = 0$, and the speed of the vehicle is

$$v = \frac{ds}{dt},$$

where $v(0) = 0$. That is, the vehicle begins its journey (as physicists put it) "from rest." Further, if D is the distance between A and B, and if T is time it takes for the entire trip, then $s(t) = D$.

Figure 19.2 again represents the Earth, with the three acceleration vectors shown. The symbol used in that figure for time derivatives is the *dot notation* introduced by the great Newton in his development of the calculus, where

$$\dot{s} = \frac{ds}{dt},$$

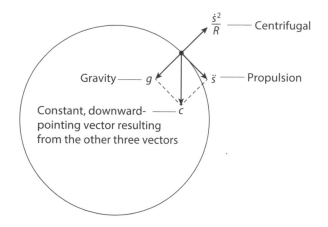

Figure 19.2. The accelerations of the transit-tube vehicle

and

$$\ddot{s} = \frac{d^2s}{dt^2} = \frac{d\dot{s}}{dt}.$$

We'll use this notational technique again in the epilogue, so pay close attention!

Writing g as the gravitational acceleration of Earth at its surface (where the transit tube is located), the net inward-directed acceleration of the vehicle is

$$g - \frac{v^2}{R} = g - \frac{1}{R}\left(\frac{ds}{dt}\right)^2 = g - \frac{1}{R}\dot{s}^2,$$

where R is the radius of the Earth. You'll notice that a fourth acceleration vector is shown in Figure 19.2, with magnitude c and pointing downward (see the end of the next paragraph for what *downward* means). This acceleration vector is the result of combining the gravitational, centrifugal, and propulsion vectors and then *defining* that resultant to be of *constant* magnitude and *always* pointing straight down. These constraints will, as you'll see next, determine $s(t)$.

Allen (see note 2) suggests a value of 40 feet/seconds-squared for the magnitude of the c-vector, a value just 25% greater than 1 gee (a 160-pound person would feel like 200 pounds). This is far less extreme than what is experienced on most roller coaster rides and should easily be tolerated by healthy people, especially for the relatively brief durations of a transit-tube trip. On long-distance trips the c-vector can rotate through a fairly large angle. So, if passenger seats are allowed to rotate on a support bar running the width of the seat (in the shoulder-to-shoulder direction, so to speak), passengers will have the sense of remaining stationary (with a constant force firmly pressing them into their seat) while the *vehicle* rotates around them. On short-distance trips (say, New York City to Boston) the rotation effect should be hardly noticeable. Passengers might even have to accept the indignity of hanging like dry cleaning on hooks, with their feet hanging downward in the direction of the c-vector, but given the state of airline travel these days, most are probably already prepared for that next (inevitable?) step in road-warrior abuse.

To start our analysis, we use the Pythagorean theorem:

$$(\ddot{s})^2 + \left(g - \frac{1}{R}\dot{s}^2\right)^2 = c^2. \qquad\qquad \text{(A)}$$

From the chain rule of calculus, where differentials are treated just like algebraic quantities (look in any freshman calculus text for details), we have

$$\ddot{s} = \frac{d\dot{s}}{dt} = \left(\frac{d\dot{s}}{ds}\right)\left(\frac{ds}{dt}\right) = \frac{d\dot{s}}{ds}\dot{s},$$

and so boxed equation (A) becomes

$$\left(\frac{d\dot{s}}{ds}\dot{s}\right)^2 + \left(g - \frac{1}{R}\dot{s}^2\right)^2 = c^2.$$

Solving for the differential ds, we have

$$ds = \frac{\dot{s}\, d\dot{s}}{\sqrt{c^2 - \left(g - \frac{1}{R}\dot{s}^2\right)^2}}.$$

Integrating *indefinitely* we obtain

$$s + k = \int \frac{\dot{s}\, d\dot{s}}{\sqrt{c^2 - \left(g - \frac{1}{R}\dot{s}^2\right)^2}},$$

where k is, for now, an arbitrary constant. We'll figure out what k actually is in just a bit. To "do" the integral on the right, we first change the integration variable to x (which changes nothing, except now we don't have to keep putting a dot over the s). That is,

$$\int \frac{\dot{s}\, d\dot{s}}{\sqrt{c^2 - \left(g - \frac{1}{R}\dot{s}^2\right)^2}} = \int \frac{x\, dx}{\sqrt{c^2 - \left(g - \frac{1}{R}x^2\right)^2}}, \quad x = \dot{s}.$$

Now, we change the variable again to

$$u = g - \frac{1}{R}x^2,$$

and so,

$$\frac{du}{dx} = -\frac{2x}{R},$$

or

$$dx = -\frac{R}{2x}\, du.$$

Thus, our integral becomes

$$\int \frac{x\, dx}{\sqrt{c^2 - \left(g - \frac{1}{R}x^2\right)^2}} = -\frac{R}{2} \int \frac{du}{\sqrt{c^2 - u^2}} = -\frac{R}{2}\sin^{-1}\left(\frac{u}{c}\right)$$

where the rightmost expression is simply looked up in a table of integrals (the easiest way to "do" integrals).[4] Now, since

$$u = g - \frac{1}{R}x^2 = g - \frac{1}{R}\dot{s}^2,$$

we have

$$s + k = -\frac{R}{2}\sin^{-1}\left(\frac{g - \frac{1}{R}\dot{s}^2}{c}\right).$$

To finish up, we have one obvious question left to answer: what's k? Well, we know that at the start of our trip through the transit tube (at $t = 0$) we have $s(0) = 0$, and that we start *from rest*. That is, $\dot{s}(0) = 0$. So, putting in these so-called initial conditions, we have

$$k = -\frac{R}{2}\sin^{-1}\left(\frac{g}{c}\right),$$

which gives us

$$s(t) = \frac{R}{2}\left[\sin^{-1}\left(\frac{g}{c}\right) - \sin^{-1}\left(\frac{g - \frac{1}{R}\dot{s}^2}{c}\right)\right]. \tag{B}$$

We're not done with our analysis yet, but just so we don't get lost in a lot of symbolic mud, let me now pause for a moment and explain what (B) is telling us. We can solve (B) for $\dot{s}(t)$, the speed of the vehicle at time t, as a function of $s(t)$, the distance traveled by time t, to arrive at

$$\dot{s}(t) = \sqrt{R}\sqrt{g - c\sin\left\{\sin^{-1}\left(\frac{g}{c}\right) - \frac{2}{R}s(t)\right\}}. \tag{C}$$

(You should convince yourself that (C) is dimensionally correct, that the right-hand side does indeed have the units of length/second.) Now, we can use this result to calculate the maximum speed of the vehicle as it travels through the transit tube, as follows.

Consider the *symmetry* of a trip. The vehicle starts at A, with $s = 0$, and then accelerates as required to achieve the net constant downward acceleration c of a passenger into his or her seat. Acceleration continues until the trip is half over, at $s = \frac{1}{2}D$, at which time the vehicle speed is maximum. The vehicle decelerates with the mirror image (negative) of the acceleration of the first half of the trip to bring the vehicle to a stop at $s = D$ at time $t = T$. (This tells us that the $s = \frac{1}{2}D$ point is reached at $t = \frac{1}{2}T$.) So, if we set $s = \frac{1}{2}D$ in (C), then $\dot{s}(t)$ will have its maximum value. Inserting the values $c = 40$ feet/seconds-squared, $g = 32.2$ feet/seconds-squared, $R = 3,960$ miles $= 2.09 \times 10^7$ feet, and $s = \frac{1}{2}D = 5,000$ miles $= 2.64 \times 10^7$ feet (New York City—Melbourne, Australia trip), gives us the result $\dot{s}_{max} = 38,830$ feet/second $= 7.35$ miles/second.

That's pretty fast, and, of course, anybody riding the transit tube would certainly find it exciting and fun (?) to know they were moving (at midpoint) at more than 26,000 miles per hour. But the questions anybody thinking of using the transit tube would ask are (1) how much does it cost? and (2) how long is the trip going to take? The first question is in the domain of economics, not physics, but we can answer the second one with just a bit more mathematics, as follows.

From boxed equation (A),

$$\left(\frac{d\dot{s}}{dt}\right)^2 = c^2 - \left(g - \frac{1}{R}\dot{s}^2\right)^2,$$

which we can solve for the differential dt to get

$$dt = \frac{d\dot{s}}{\sqrt{c^2 - \left(g - \frac{1}{R}\dot{s}^2\right)^2}}.$$

Formal integration gives

$$\int dt = \int \frac{d\dot{s}}{\sqrt{\left\{c - \left(g - \frac{1}{R}\dot{s}^2\right)\right\}\left\{c + \left(g - \frac{1}{R}\dot{s}^2\right)\right\}}}.$$

Notice that the integration limits have been omitted on both integrals; we'll determine them shortly. Now, again using our earlier simplifying change of notation $x = \dot{s}$, we get

$$\int dt = \int \frac{dx}{\sqrt{\{c - (g - \frac{1}{R}x^2)\}\{c + (g - \frac{1}{R}x^2)\}}}, \quad x = \dot{s}. \tag{D}$$

We next make the change of variable from x to ϕ, where we define their relationship to be

$$x = \sqrt{R(c + g)}\cos(\phi),$$

a definition that must seem to be quite mysterious. We are certainly free to make any change we wish, but why *this* one? The easy answer is that it will eventually give us a known integral, but the question still remains: how does one know ahead of time that *this* change "works"? Ah, that's what makes the reputation of the analyst (not mine, I hasten to add, as this change is from Allen's paper of note 2)!

In any case, continuing, we have

$$g - \frac{1}{R}x^2 = g - \frac{R(c + g)}{R}\cos^2(\phi) = g - (c + g)\cos^2(\phi)$$

$$= g\sin^2(\phi) - c\cos^2(\phi).$$

Also,

$$\frac{dx}{d\phi} = -\sqrt{R(c + g)}\sin(\phi),$$

and so, returning to boxed expression (D) and doing a little algebra, we obtain

$$\int dt = -\sqrt{R(c + g)}$$

$$\times \int \frac{\sin(\phi)\,d\phi}{\sqrt{\{c - g\sin^2(\phi) + c\cos^2(\phi)\}\{c + g\sin^2(\phi) - c\cos^2(\phi)\}}}$$

$$= -\sqrt{R(c+g)}$$

$$\times \int \frac{\sin(\phi)\,d\phi}{\sqrt{\left\{c\left[1+\cos^2(\phi)\right]-g\sin^2(\phi)\right\}\left\{c\left[1-\cos^2(\phi)\right]+g\sin^2(\phi)\right\}}}$$

$$= -\sqrt{R(c+g)}$$

$$\times \int \frac{\sin(\phi)\,d\phi}{\sqrt{\left\{c\left[2-\sin^2(\phi)\right]-g\sin^2(\phi)\right\}\left\{c\sin^2(\phi)+g\sin^2(\phi)\right\}}}$$

$$= -\sqrt{R(c+g)}\int \frac{\sin(\phi)\,d\phi}{\sqrt{\left\{2c-c\sin^2(\phi)-g\sin^2(\phi)\right\}(c+g)\sin^2(\phi)}}$$

$$= -\sqrt{R}\int \frac{d\phi}{\sqrt{2c-(c+g)\sin^2(\phi)}} = -\frac{\sqrt{R}}{\sqrt{2c}}\int \frac{d\phi}{\sqrt{1-\left(\frac{c+g}{2c}\right)\sin^2(\phi)}}.$$

So, defining the constant k^2, we have

$$\int dt = -\sqrt{\frac{R}{2c}}\int \frac{d\phi}{\sqrt{1-k^2\sin^2(\phi)}}, \qquad k^2 = \frac{c+g}{2c}.$$

At this point we can no longer avoid the issue of integration limits. Here's how to get them for both integrals. When $t=0$ we know $\dot{s}=0$, and so, since $x=\dot{s}$, we know from $x=\sqrt{R(c+g)}\cos(\phi)$ that at $t=0$ we have $\cos(\phi)=0$. That is, $\phi=\frac{\pi}{2}$ when $t=0$. Now, what's ϕ equal to when the trip is half over, that is, when $t=\frac{T}{2}$? Let's call that value ϕ_1. From our earlier work we know that $\dot{s}$ is at its maximum value at that time, and so

$$\dot{s}_{\max} = \sqrt{R(c+g)}\cos(\phi),$$

which says[5] that

$$\phi_1 = \cos^{-1}\left\{\sqrt{\frac{\dot{s}^2_{\max}}{R(c+g)}}\right\}.$$

If we square both sides of the $\dot{s}_{max}$ equation, using the boxed expression (C) with $s = \frac{D}{2}$ inserted because that's where $\dot{s}_{max}$ occurs, we have

$$\dot{s}^2_{max} = R\left[g - c\sin\left\{\sin^{-1}\left(\frac{g}{c}\right) - \frac{D}{R}\right\}\right].$$

So, putting in limits, we get

$$\int_0^{T/2} dt = \frac{T}{2} = -\sqrt{\frac{R}{2c}}\int_{\pi/2}^{\phi_1}\frac{d\phi}{\sqrt{1 - k^2\sin^2(\phi)}},$$

or, at last,

$$
T = \sqrt{\frac{2R}{c}}\left[\int_0^{\pi/2}\frac{d\phi}{\sqrt{1 - k^2\sin^2(\phi)}} - \int_0^{\phi_1}\frac{d\phi}{\sqrt{1 - k^2\sin^2(\phi)}}\right],
$$

$$
k^2 = \frac{c + g}{2c},
$$

$$
\dot{s}^2_{max} = R\left[g - c\sin\left\{\sin^{-1}\left(\frac{g}{c}\right) - \frac{D}{R}\right\}\right],
$$

$$
\phi_1 = \sin^{-1}\left\{\sqrt{1 - \frac{\dot{s}^2_{max}}{R(c + g)}}\right\}.
$$

(E)

For $R = 2.09 \times 10^7$ feet, $c = 40$ feet/seconds-squared, and $g = 32.2$ feet/seconds-squared we find that $\sqrt{\frac{2R}{c}} = 1,022$ seconds and that $k^2 = 0.9025$ ($k = 0.95$). Also, as we found earlier, for $D = 10,000$ miles (the Melbourne, Australia/New York City trip) $\dot{s}_{max} = 38,830$ feet/second, and so $\phi_1 \approx 0$. Now, both of the integrals in (E) are what mathematicians call *elliptic integrals of the first kind*, and they are an entirely new function (of two parameters, k and the upper-limit angle). They cannot be expressed in terms of the "ordinary" functions of math, like exponentials, trig functions, and square roots (or other powers). They have to be numerically calculated (which you then look up in tables) or evaluated by coded algorithms on a computer as you need them. I used a Web-based calculator[6] available for free, and obtained the result $T = 1,022(2.59 - 0)$ seconds $= 2,647$ seconds $= 44.1$ minutes. That sure beats 20 hours crammed into a jet airplane seat in coach, with the reclined guy in front of you sleeping (maybe even snoring) in your lap.

Okay, this has all been fun (hasn't it?), but who is *really* going to build a great circle transit tube between New York City and Melbourne, Australia? Think of that *really deep ocean* between those two cities, in which there would have to be some pretty beefy support towers! More likely are transit tubes connecting New York City, Boston, and Washington, DC, as those tubes would be entirely over land, and erecting support towers (or digging shallow, below-surface tunnels) for the tubes would at least be feasible tasks.

Many people also regularly travel between the two U.S. coasts, and I'll let you verify that a transit-tube trip between, for example, New York City and Los Angeles (2,450 miles) would require just 23.3 minutes and would reach a top speed of 3.84 miles/second, in comparison with more than 300 minutes by commercial jet. And if you really like to punch numbers into your calculator, here are four more examples to check.

TABLE 19.1
Example Transit-Tube Times.

A	B	D (miles)	$\dot{s}_{max}$ (miles/second)	T (minutes)
Minsk	Beijing	1,814	3.22	20.3
Paris	Moscow	1,550	2.94	18.8
Paris	Berlin	546	1.64	11.3
London	Paris	213	0.99	6.9

Russian and Chinese tourists should like the first entry a lot. And the last entry is particularly impressive when compared with taking the

Eurostar train, which requires 135 minutes for the London/Paris run. With a transit tube, however, you can be in London at 10:00 am and in Paris before 10:07 am. Very nifty, indeed!

As a final "assignment," you might find it interesting to compare the transit tube with a proposed high-speed tube system linking San Francisco and Los Angeles, the so-called Hyperloop.[7]

Notes

1. You can find a mathematical discussion on such tunnels (with historical commentary) in my book *Mrs. Perkins's Electric Quilt*, Princeton University Press, 2009, pp. 203–214.

2. William A. Allen, "Two Ballistic Problems for Future Transportation," *American Journal of Physics*, February 1953, pp. 83–89. This is a difficult paper for beginning students to read. It contains equations with terms like $\int_0^s f(s)\,ds$, and while an experienced analyst will know what is meant, to a beginning calculus student it will be nonsense (having the integration variable s vary from zero to itself is, on its face, meaningless). The presentation I give here is both an expanded version of Allen's presentation and a somewhat different mathematical treatment. The final results, however, are the same.

3. The only exception to this statement is if the two points are the endpoints of a diameter of the sphere, and then there are an *infinite* number of great circle paths, all equal in length to half the length of an equatorial circumference.

4. This "method" depends, of course, on having a table of integrals that contains an entry for the particular integral in which you are interested. If not, you'll have to do the integral yourself. "Doing" integrals has a long and eventful history in mathematics: see, for example, my book *Inside Interesting Integrals*, Springer, 2015, and George Boros and Victor Moll, *Irresistible Integrals*, Cambridge University Press, 2004.

5. In Allen's paper $\phi_1 = \sin^{-1}\left\{\sqrt{1 - \frac{\dot{s}_{max}^2}{R(c+g)}}\right\}$, but it is easy to show that these two expressions are equivalent. (*Hint*: Draw a right triangle with ϕ_1 as one of the acute angles and then apply the Pythagorean theorem and the definitions of the sine and the cosine.) Allen's expression is preferable, however, as in situations where ϕ_1 is very close to zero it is guaranteed not to fail because of round-off noise (the inverse-cosine form can result in an argument slightly *greater* than 1, which produces an error).

6. At keisan.casio.com/exec/system/1244989500. Elliptic integrals occur all over the place in advanced physics and, as we've seen here, in simple physics, too. (You can find more discussion on their appearance in physics in *Inside Interesting Integrals* (see note 4), pp. 212–219.) In the final chapter of this book I'll show you yet another occurrence of an elliptic integral, in an even simpler situation than the transit-tube problem.

7. James Vlahos, "Hyped Up," *Popular Science*, July 2015, pp. 32–39.

20. Hurtling Your Body through Space

A coward turns away, but a brave man's choice is danger.
— *Euripides (ca. 400 BC), who might have added that the brave
often die young*

People have always done all sorts of foolish things, from tossing spears at woolly mammoths and then running like crazy to escape the enraged beasts, to jumping off 500-foot-high bridges with thin elastic cords tied to their ankles to jerk them to a stop after falling 499 (instead of 501) feet. In this chapter, in addition to bungee-cord jumping, we'll discuss two other only somewhat less dangerous yet quite common human activities that involve hurtling through space: the "Tarzan" swing-and-release from a hanging rope to zoom over a dank swamp infested with poisonous snakes (Indiana Jones would be particularly interested in our results for this situation!) and the ski jump. In all three of our analyses we'll use a lot of the simple physics that we've already developed earlier in the book, plus some new stuff we haven't seen before.

The Ski Jump

This problem, the simplest of the three, is described by the geometry shown in Figure 20.1. A skier accelerates down a takeoff ramp, which is constructed with a little upward curve at the end so that the skier leaves the end of the ramp (at the origin of our coordinate system) at an angle α at some speed v_0. The skier then hurtles through space on

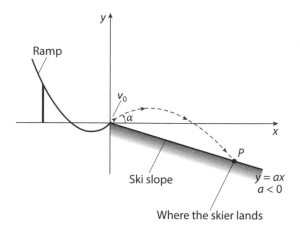

Figure 20.1. The geometry of a ski jump

a parabolic path (refer to Figure 18.1) until landing on a ski slope at point P. The ski slope starts at the origin and has negative slope a, as shown in the figure. Our problem is to determine what α should be to maximize the length of the jump (that is, we want to make the x-coordinate of P as large as possible).[1] We will ignore any air drag effects and consider gravity to be the only force at play.

You'll recall that in Chapter 18 we derived the equation for the parabolic path of a projectile leaving the origin at angle α at speed v_0:

$$y = x \tan(\alpha) - \frac{g}{2v_0^2 \cos^2(\alpha)} x^2.$$

Our skier lands on the ski slope (with equation $y = ax$) at P, and so the x-coordinate of P satisfies

$$ax = x \tan(\alpha) - \frac{g}{2v_0^2 \cos^2(\alpha)} x^2,$$

or

$$\frac{g}{2v_0^2 \cos^2(\alpha)} x^2 = x[\tan(\alpha) - a].$$

In addition to the trivial solution $x = 0$, we have

$$x = \frac{2v_0^2 \cos^2(\alpha)[\tan(\alpha) - a]}{g} = \frac{2v_0^2 \cos^2(\alpha)\left[\frac{\sin(\alpha)}{\cos(\alpha)} - a\right]}{g}$$

$$= \frac{2v_0^2}{g}\left[\cos(\alpha)\sin(\alpha) - a\cos^2(\alpha)\right].$$

Thus,

$$\frac{dx}{d\alpha} = \frac{2v_0^2}{g}\left[\{\cos^2(\alpha) - \sin^2(\alpha)\} + a\{2\cos(\alpha)\sin(\alpha)\}\right].$$

Recalling that the expression in the first pair of curly brackets is $\cos(2\alpha)$, and that the expression in the second pair of curly brackets is $\sin(2\alpha)$, we have

$$\frac{dx}{d\alpha} = \frac{2v_0^2}{g}[\cos(2\alpha) + a\sin(2\alpha)].$$

Setting this equation equal to zero to maximize x, we obtain

$$\frac{\sin(2\alpha)}{\cos(2\alpha)} = \tan(2\alpha) = -\frac{1}{a}, \quad a < 0.$$

So, if $a = 0$ (the "slope" isn't actually a slope but is, instead, horizontal), then we get the expected $\alpha = 45°$ for the maximum-distance jump, while if $a = -1$ (giving a steep 45° slope) the maximum-distance jump occurs for $\alpha = 22.5°$. Notice that this result, for the optimal α, is *independent* of v_0 (it is the best angle for *all* skiers of *any* strength who use this particular jump facility) and is a function of only the steepness of the landing slope. The steeper the slope, the smaller α should be.

In the limit $a = -\infty$ (which means the "slope" is actually a *vertical cliff*, we have $\alpha = 0$, which means the skier shoots horizontally straight off the ramp parallel to the x-axis. Physically, the skier never hits the "slope" but simply keeps moving forward, all the while falling vertically. If you plug $\alpha = 0$ and $a = -\infty$ into the equation for the x-coordinate

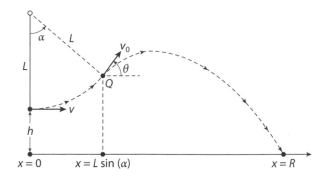

Figure 20.2. The geometry of a Tarzan swing

of point P, you get $x = \infty$. This is all theoretical, however; in the real world, a skier ends up at the bottom of a very deep ravine.

The Tarzan Swing

In this problem we have a man—Tarzan or Indiana Jones, for example—modeled as a point mass m, running toward a vine of length L that is hanging straight down from an overhead tree branch. The point over which the lower end of the vine hangs we'll call the origin of an x, y-coordinate system, as shown in Figure 20.2. The lower end of the vine is distance h above $x = 0$. When the man reaches $x = 0$ he is moving at speed v, and at that instant, he grabs the end of the vine and so swings outward and upward, like the bob at the end of a pendulum. At some point, Q, when the vine has rotated through angle α, he lets go of the vine and then arcs through space on a parabolic path until he arrives on the x-axis at $x = R$. Our question is simple: what release angle α maximizes R?

In the following analysis we'll ignore air drag (as in the ski jump problem) and we'll assume the vine swings from its tree branch without friction. We'll write v_0 as the speed of the man at the point of release, Q, and use the geometric fact that the angle α through which the vine has rotated at the point of release is the angle θ the man's velocity vector makes with the horizontal. Figure 20.3 illustrates this situation,

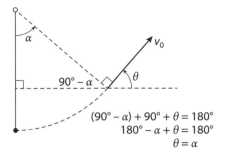

Figure 20.3. Why $\theta = \alpha$ in Figure 20.2

which follows from the observation that, at release, the velocity vector of the man is *perpendicular* to the vine.

To start our analysis,[2] at the instant the man reaches $x = 0$ and grabs the vine he has zero potential energy and $\frac{1}{2}mv^2$ kinetic energy. If the vine then swings through angle α, the man rises vertically through the distance

$$L - L\cos(\alpha)$$

and so gains potential energy

$$mgL\{1 - \cos(\alpha)\}.$$

This energy comes from his kinetic energy and so, when the vine has swung through angle α, the man is moving at speed v_0 and has kinetic energy

$$\frac{1}{2}mv_0^2 = \frac{1}{2}mv^2 - mgL\{1 - \cos(\alpha)\}.$$

If the man releases his grip on the vine at this point (Q), then his speed as he starts his parabolic arc through space is

$$v_0 = \sqrt{v^2 - 2gL\{1 - \cos(\alpha)\}}$$

at angle α to the horizontal. His coordinates at Q, at that instant, are $x = L\sin(\alpha)$ and $y = L\{1 - \cos(\alpha)\}$.

It is clear that depending on the speed of the running man at the instant he grabs the vine, the angle through which the vine can swing, *at most*, is given by

$$\alpha_{max} = \cos^{-1}\left\{1 - \frac{v^2}{2gL}\right\}.$$

This is the angle at which all his kinetic energy has been converted to potential energy. The largest possible α of physical interest is, of course, $90°$,[3] and so if $v^2 > 2gL$, then all values for the release angle that are of physical interest are possible. If $v^2 < 2gL$ then the values for the release angle that are of physical interest are in the interval $0 < \alpha \leq \alpha_{max}$. For a 20-foot vine, for example, the critical speed v that separates these two cases is

$$v = \sqrt{2 \times 32.2 \times 20}\,\text{feet/second} \approx 36\,\text{feet/second}.$$

This *is* pretty fast, corresponding to running a 100-yard dash in less than 8.4 seconds, more than a second faster (as I write) than the world record! A more reasonable way for the man to reach this speed is to simply do what Tarzan is famous for—instead of running, imagine he is initially high up in a tree on a platform, and grasping a fully stretched vine, he launches himself into space. When he passes over the origin he can *easily* be moving at 36 feet/second or more.

Now, put all that aside for the moment and again recall the equation for the parabolic path of a projectile leaving the origin at angle α at speed v_0:

$$y = x\tan(\alpha) - \frac{g}{2v_0^2\cos^2(\alpha)}x^2.$$

We used this equation in this form in the ski jump analysis, but here things are just a bit different. The angle α and the speed v_0, are as before, but now the man is *not* leaving the origin but, rather, is distance

$$h = L\{1 - \cos(\alpha)\}$$

above the x-axis when he lets go of the vine. This is easy to correct for, however. Imagine the man *is* leaving the origin, and we now ask what

x is when $y = -h$. That is, let's solve

$$-h = x \tan(\alpha) - \frac{g}{2v_0^2 \cos^2(\alpha)} x^2$$

for x. This is easily done with the quadratic equation formula, and I'll let you verify that the answer is

$$x = \frac{v_0^2}{2g} \left[\sin(2\alpha) + \sqrt{\sin^2(2\alpha) + \frac{8hg}{v_0^2} \cos^2(\alpha)} \right].$$

Be sure that the physical significance of this x is clear in your mind—it is the distance between the x-coordinate of Q and $x = R$. So, to get the value of R, itself, we must add the horizontal distance (the x-coordinate of Q) the man swings across on the vine before he lets go:

$$R = L \sin(\alpha) + \frac{v_0^2}{2g} \left[\sin(2\alpha) + \sqrt{\sin^2(2\alpha) + \frac{8hg}{v_0^2} \cos^2(\alpha)} \right],$$

$$v_0 = \sqrt{v^2 - 2gL \{1 - \cos(\alpha)\}}.$$

To find the release angle α that maximizes R, a pure mathematician might say, "no problem, just set $\frac{dR}{d\alpha} = 0$ and solve for α." Well, you can do that—*if* you're a glutton for agony—but I'm going to take a different approach and use a computer to plot R versus α and simply *see* where R peaks. To make our numerical work as useful as possible, however, let's first normalize the R equation into one involving dimensionless variables (recall that we used this method in Chapter 16) Following the lead of a recent paper,[4] we use a "natural" length L (the length of the vine) and a "natural" speed $\sqrt{2gL}$ (examine the expression we derived for v_0). So, we define the variables

$$w = \frac{v}{\sqrt{2gL}}, \quad s = \frac{h}{L}, \quad a = \cos(\alpha).$$

Then,

$$\frac{R}{L} = \sin(\alpha) + \frac{v_0^2}{2gL}\left[\sin(2\alpha) + \sqrt{\sin^2(2\alpha) + \frac{8hg}{v_0^2}\cos^2(\alpha)}\,\right],$$

and, since (as you can verify)

$$\sin(\alpha) = \sqrt{1-a^2}, \quad \sin(2\alpha) = 2\sin(\alpha)\cos(\alpha) = 2a\sqrt{\left(1-a^2\right)},$$

$$\frac{v_0^2}{2gL} = w^2 - 1 + a, \quad \frac{8hg}{v_0^2}\cos^2(\alpha) = \frac{4sa^2}{w^2-1+a},$$

we have

$$\frac{R}{L} = \sqrt{1-a^2} + 2a\left(w^2 - 1 + a\right)\left[\sqrt{1-a^2} + \sqrt{\left(1-a^2\right) + \frac{s}{w^2-1+a}}\,\right].$$

A "typical" value for s might be $\frac{1}{3}$ (for example, a 15-foot vine with its bottom end 5 feet above ground at the low point of the swing). For a 15-foot vine,

$$\sqrt{2gL} = \sqrt{2 \times 32.2 \times 15}\,\text{feet/second} = 31\,\text{feet/second}.$$

So, if we pick $w = 1$, then 0.7, and then 0.4, we'll have Tarzan's speed when he grabs the vine as 31 feet/second, 21.7 feet/second, and 12.4 feet/second, respectively. Figure 20.4 shows three plots of $\frac{R}{L}$ versus α as α varies from 0 to $\alpha_{\max}$: the top plot is for $w = 1$, the middle plot is for $w = 0.7$, and the bottom plot is for $w = 0.4$.

In Figure 20.4 we see that each plot does have a well-defined peak (the peaks are broad, however, showing that the particular α Tarzan uses isn't critical for getting him across the swamp) and that the angle that gives the maximum range increases as w (his speed when he grabs the vine) increases. In all the plots the maximum range occurs at a launch angle considerably less than $45°$; for $w = 0.7$, for example, the optimal launch angle is only about $30°$.

Now, for those readers who are wondering if this is really a problem of everyday life (after all, how many swamps have *you* swung over in the

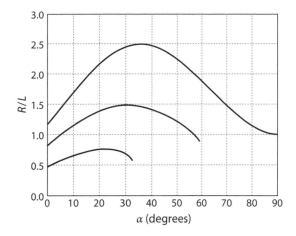

Figure 20.4. Normalized range versus the launch angle

last 10 years?), let me remind you that probably every one of you has done a Tarzan swing at one time or another. Just think back to when you were a kid at a playground, on a swing, pumping it higher and higher. And then, on the last forward swing before you had to get home for dinner, you launched yourself out of the swing and landed in the surrounding sandpit. Remember doing that? *That was a Tarzan swing*!

The Bungee Jump

A daredevil (modeled as a point mass m) ties one end of a long, elastic, *massless* cord to an ankle, and the other end to the edge of a bridge column several hundreds of feet above a rocky gorge, and steps off into space. As the jumper falls the cord trails out behind him until he has fallen a distance equal to the length of the cord, L_0. He continues to fall because the cord begins to stretch; we'll assume the cord obeys *Hooke's law*[5] as it stretches. That is, if we label the vertical axis as y and positive-increasing in the *downward* direction (see Figure 20.5) and set $y = 0$ (at time $t = 0$) as the point where the cord stretching just begins, then the tension in the cord (directed upward in the *negative* y-direction *toward* the bridge) is, for k some positive constant, given by ky. This force, which is in the direction opposite that of gravity,

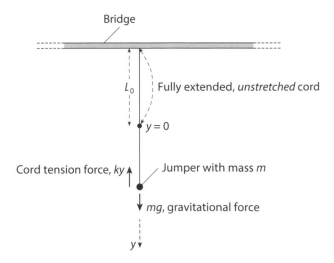

Figure 20.5. The geometry of a bungee jump

slows the fall, eventually brings the jumper to a stop, and then pulls him upward.

There's a double thrill in doing this crazy (to me) stunt: not being smashed on the rocks below, and experiencing an acceleration greater than that of gravity. Indeed, that's the question we'll answer here with some simple physics—what's the maximum acceleration experienced by the jumper? As you'll see, it can be considerably greater than 1 gee.

As the jumper falls and until the cord begins to stretch, the only force he feels is the downward-directed gravitational force mg. Once the cord begins to stretch, however, he also feels the tension force, directed upward, of ky, and so the total force on the jumper for $y \geq 0$ is

$$F = m \frac{d^2 y}{dt^2} = mg - ky,$$

and so,

$$\frac{d^2 y}{dt^2} + \frac{k}{m} y = g, \quad y \geq 0.$$

This is a common, very important differential equation that engineers, physicists, and mathematicians often encounter, and its solution is well known to them. It's perhaps a bit beyond what most high school math programs teach, however, and so I'll spend a little time showing you how to solve it—it's really not that difficult to do.

Let's start by *assuming* the general form of the solution is the sum of a constant and a time-varying function. That would certainly seem to cover a lot of bases! If C is the constant, then plugging $y = C$ into the differential equation gives

$$\frac{k}{m} C = g,$$

and we have our constant:

$$C = \frac{mg}{k}.$$

So, writing the time-varying part of the solution as $f(t)$ gives us the total solution

$$y(t) = \frac{mg}{k} + f(t).$$

If we next plug this expression into the differential equation, we get

$$\frac{d^2 f}{dt^2} + \frac{k}{m} \left[\frac{mg}{k} + f(t) \right] = g,$$

or

$$\frac{d^2 f}{dt^2} + \frac{k}{m} f(t) = 0,$$

and so we have a differential equation for just $f(t)$:

$$\frac{d^2 f}{dt^2} = -\frac{k}{m} f(t).$$

That is, $f(t)$ is a function such that its *second* derivative is a scaled version of itself. Can you think of functions that have that property? Sure you can—sines and cosines![6]

So, suppose (with A and ω constants) that

$$f(t) = A\cos(\omega t).$$

Then,

$$\frac{d^2 f}{dt^2} = -A\omega^2 \cos(\omega t)$$

and, substituting this expression into the differential equation for $f(t)$, we get

$$-A\omega^2 \cos(\omega t) = -\frac{k}{m} A \cos(\omega t),$$

which says

$$\omega^2 = \frac{k}{m}.$$

We can also suppose that

$$f(t) = B\sin(\omega t),$$

and so, again, we are led to

$$\omega^2 = \frac{k}{m}.$$

Thus, most generally, we can write

$$f(t) = A\cos(\omega t) + B\sin(\omega t), \quad \omega = \sqrt{\frac{k}{m}},$$

and our complete solution for $y(t)$ is

$$y(t) = \frac{mg}{k} + A\cos(\omega t) + B\sin(\omega t), \quad \omega = \sqrt{\frac{k}{m}}.$$

Now, what are A and B? Well, we can find A because we know that at $t = 0$ we have $y(0) = 0$, and so

$$0 = \frac{mg}{k} + A,$$

which says that

$$A = -\frac{mg}{k}.$$

Thus,

$$y(t) = \frac{mg}{k} - \frac{mg}{k} \cos(\omega t) + B \sin(\omega t),$$

or

$$y(t) = B \sin(\omega t) + \frac{mg}{k}[1 - \cos(\omega t)].$$

To find B we write

$$\frac{dy}{dt} = B\omega \cos(\omega t) + \frac{mg}{k}\omega \sin(\omega t).$$

We can use this expression by observing that if the speed of the jumper, just as the cord begins to stretch at time $t = 0$ is v_0, then

$$\frac{dy}{dt}\Big|_{t=0} = v_0 = B\omega,$$

and so

$$B = \frac{v_0}{\omega},$$

which gives us

$$y(t) = \frac{v_0}{\omega} \sin(\omega t) + \frac{mg}{k}[1 - \cos(\omega t)].$$

We can find v_0 by recognizing that if the jumper falls through distance L_0 in a time interval T, then[7] $\frac{1}{2}gT^2 = L_0$, and so $T = \sqrt{\frac{2L_0}{g}}$. The speed at the end of time interval T is

$$gT = g\sqrt{\frac{2L_0}{g}} = \sqrt{2g L_0} = v_0.$$

The acceleration of the jumper for $y > 0$ is

$$\frac{d^2 y}{dt^2} = \frac{v_0}{\omega}\{-\omega^2 \cos(\omega t)\} - \frac{mg}{k}\omega^2 \sin(\omega t) = -v_0\omega \cos(\omega t) - \frac{mg}{k}\omega^2 \sin(\omega t)$$

$$= -\sqrt{2g L_0}\sqrt{\frac{k}{m}} \cos(\omega t) - \frac{mg}{k}\left(\frac{k}{m}\right)\sin(\omega t)$$

$$= -\left[\sqrt{\frac{2g L_0 k}{m}} \cos(\omega t) + g \sin(\omega t)\right].$$

This acceleration is of the general form

$$\frac{d^2 y}{dt^2} = a \cos(\omega t) + b \sin(\omega t),$$

with

$$a = -\sqrt{\frac{2g L_0 k}{m}}, \quad b = -g.$$

I'll let you show, using the hint in note 8,[8] that the magnitude of the maximum acceleration is

$$\max\left|\frac{d^2 y}{dt^2}\right| = \sqrt{a^2 + b^2} = \sqrt{\frac{2g L_0 k}{m} + g^2} = g\sqrt{1 + \frac{2L_0 k}{mg}}.$$

Now, to finish this analysis, and to put our result for the maximum acceleration in easy-to-appreciate form, let's take a more detailed look at the constant k. Since the tension force in the stretching cord for $y > 0$ is $F = ky$, the *stretching* elastic cord is storing energy E. Let the maximum length of the cord during the jump be L_m. Then, since the cord *stretches* by the amount $L_m - L_0$, the energy in the *stretched* cord is

$$E = \int_0^{L_m - L_0} F\,dy = \int_0^{L_m - L_0} ky\,dy = \frac{1}{2}ky^2|_0^{L_m - L_0} = \frac{1}{2}k\,(L_m - L_0)^2.$$

This energy comes from a decrease, as he falls, of the jumper's potential energy. Since he falls distance L_m from the point the cord *starts to stretch*, the decrease in potential energy is $mg\,L_m$, and so

$$\frac{1}{2}k\,(L_m - L_0)^2 = mg\,L_m.$$

Thus,

$$k = \frac{2mg\,L_m}{(L_m - L_0)^2},$$

and so

$$\max\left|\frac{d^2y}{dt^2}\right| = g\sqrt{1 + \frac{2L_0\frac{2mg\,L_m}{(L_m - L_0)^2}}{mg}} = g\sqrt{1 + \frac{4L_m L_0}{(L_m - L_0)^2}}$$

$$= g\frac{\sqrt{(L_m - L_0)^2 + 4L_m L_0}}{L_m - L_0} = g\frac{\sqrt{L_m^2 + 2L_m L_0 + L_0^2}}{L_m - L_0}$$

$$= g\frac{L_m + L_0}{L_m - L_0},$$

or, at last,

$$\max\left|\frac{d^2y}{dt^2}\right| = g\frac{\frac{L_m}{L_0} + 1}{\frac{L_m}{L_0} - 1}.$$

This amazingly simple-looking result is quite revealing. If $L_m = 2L_0$, that is, if the cord stretches to twice its unstretched length during the jump, then

$$\max\left|\frac{d^2y}{dt^2}\right| = 3g,$$

while if the cord stretches by only 50%, that is if $L_m = \frac{3}{2}L_0$, then

$$\max \left| \frac{d^2y}{dt^2} \right| = 5g.$$

The less stretch, the greater is the maximum acceleration. In the limit of *no stretch at all* ($L_m = L_0$), we get a horrifying result. Suppose, to illustrate this, that our bungee jumper accidently ties a steel-linked chain to his ankle instead of an elastic cord. Then,

$$\max \left| \frac{d^2y}{dt^2} \right| = \infty,$$

which is the math describing the *really big* jerk to a very sudden, *dead* (literally) stop that our jumper will experience at the instant the chain becomes fully extended.

Now, some final words on the analysis I've just taken you through. First, what I've described was motivated by a challenge problem in the *American Journal of Physics*.[9] Second, a few years later a nice article[10] appeared in *The Physics Teacher* that gently tweaked the author of the *AJP* problem as having made an error—a criticism I think unwarranted. I won't go into details here, but the *AJP* problem specifically states that the cord is massless and implies that when the jump starts the cord is coiled next to the jumper *on the bridge*. The *PT* analysis, in contrast, specifically states that the cord is massive and, at the start of the jump, is hanging in a loop extending halfway down from the jumper on the bridge and then back up to the bridge. The two analyses are *both* correct but of quite different physical situations.[11]

Even with "simple physics," professional physicists can find reasons to disagree. That's one of the features of physics that makes it so interesting.

Notes

1. The analysis here is only a *very* slightly modified version of the one done by Krzysztof Rebilus, "Optimal Ski Jump," *The Physics Teacher*, February 2013, pp. 108–109.

2. Inspired after reading an elegant paper by David Bittel (a high school physics teacher in Connecticut), "Maximizing the Range of a Projectile Launched by a Simple Pendulum," *The Physics Teacher*, February 2005, pp. 98–100.

3. This angle launches Tarzan *straight up*, and therefore he'll eventually fall *straight down*. So, this angle, while interesting, is probably *not* really very helpful in swinging over a swamp (angles greater than 90° will launch Tarzan *backward*).

4. Carl E. Mungan, "Analytically Solving Tarzan's Dilemma," *The Physics Teacher*, January 2014, p. 6. Mungan elaborates on how to find the optimal launch angle by solving a particular cubic equation, an observation made earlier by Bittel (note 2).

5. Named after Newton's contemporary Robert Hooke (1635–1703), who was on what we'd today call Newton's "unfriend" list. You can read about Hooke and Newton, and of their contentious relationship, in my book *Mrs. Perkins's Electric Quilt*, Princeton University Press, 2009, pp. 167–168, 170–172, 184, 188, 190–191.

6. More generally, the *exponential* e^{st}, where s is a constant (*every* derivative of e^{st} is a scaled version of e^{st}), but to go that route will quickly lead us into exponentials with imaginary exponents. That is, in fact, the mathematically best and most general way to go in solving differential equations like ours, but for the very simple case we have here it is far more powerful than we need. Working with sines and cosines will do the job.

7. Assuming the jumper starts his fall at zero speed. That is, he simply tips forward off the bridge.

8. To show that the maximum of $f(t) = a\cos(\omega t) + b\sin(\omega t)$ is $\sqrt{a^2 + b^2}$, begin by setting $\frac{df}{dt} = 0$ and show this occurs when $t = \frac{1}{\omega}\tan^{-1}\left(\frac{b}{a}\right)$. Then, insert this t into $f(t)$ to show that $f\left\{\frac{1}{\omega}\tan^{-1}\left(\frac{b}{a}\right)\right\} = a\cos\left\{\tan^{-1}\left(\frac{b}{a}\right)\right\} + b\sin\left\{\tan^{-1}\left(\frac{b}{a}\right)\right\} = \sqrt{a^2 + b^2}$. Drawing the obvious right triangle might help in this last step.

9. Peter Palffy-Muhoray, "Acceleration during Bungee-Cord Jumping," *American Journal of Physics*, April 1993, pp. 379, 381. I've corrected a math typo in the *AJP* printing, and have greatly elaborated on how to solve the jumper's differential equation of motion, but my presentation here is essentially that of Palffy-Muhoray.

10. David Kagan and Alan Kott, "The Greater-Than-*g* Acceleration of a Bungee Jumper," *The Physics Teacher*, September 1996, pp. 368–373.

11. You can find more discussion on the mathematical physics of the *PT* situation (at a somewhat higher math level than in this book) in my book *Inside Interesting Integrals*, Springer, 2015, pp. 212–219.

21. The Path of a Punt

The mark of a great punter is a long hang time.
— *anonymous football fan, speaking a profound truth*

"Hang time" isn't a reference to the theme of Clint Eastwood's terrific 1968 Western *Hang 'Em High* but, rather, is how long it takes for a football to travel on its parabolic path[1] from the punter's foot to its catch by the receiving team. A long hang time gives the kicking team time to get downfield before the receiving team has a chance to execute a runback of the ball. Hang time *could* also have an important role in baseball, where a towering flyball deep to the outfield would give any runners already on base a longer time to get to the next base—but it doesn't. That's because of the "tag up" rule that requires base runners to retouch or remain on their starting base until (after) the ball either lands in fair territory or is first touched by a fielder. Base runners must tag up when a fly ball is caught in flight by a fielder.

It's easy to calculate the hang time for a football on a parabolic path. As we found in Chapter 18, the equation for the height of a football leaving the punter's foot at speed v_0 at angle α is, at time t, given by

$$y(t) = v_0 t \sin(\alpha) - \frac{1}{2}gt^2.$$

This relation tells us that $y(t) = 0$ when $t = 0$ (when the ball leaves the punter's foot) and at

$$t = \frac{2v_0 \sin(\alpha)}{g} = T$$

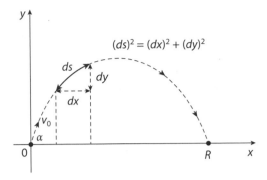

Figure 21.1. A differential portion of the parabolic path

(when the ball is caught). T is the hang time, and we see that it continuously increases as α increases from 0 to 90°. The angle α is the only parameter the punter controls, as we'll assume the speed v_0 is a function of leg strength[2] (and g is, of course, the acceleration of gravity). The longest hang time is (somewhat ironically) for $\alpha = 90°$, a *straight-up* kick, which is, of course, the *last* thing a punter wants! That's a kick that doesn't go anywhere (it has zero range), while the punter wants a *long* range to his kick. So, in fact, there's his quandary. How should he kick the ball (what should α be?) to get a long hang time *and* a long range?

One answer is to pick α so that the football travels the *longest path* on its journey through the air. This choice does, in fact, give a hang time and a range that are both significant fractions of the maximum possible for each, individually (you'll see this when we finish our analysis). So, what launch angle α gives the longest path length?

As shown in Figure 21.1, if we look at a *tiny*, arbitrary section of the parabolic path, its length (the *differential ds*, because it's *tiny*) is given by the Pythagorean theorem as

$$(ds)^2 = (dx)^2 + (dy)^2,$$

and so the total length of the path, from start to finish, is

$$L = \int_{\text{start}}^{\text{finish}} ds = \int_{\text{start}}^{\text{finish}} \sqrt{(dx)^2 + (dy)^2} = \int_{\text{start}}^{\text{finish}} \sqrt{1 + \left(\frac{dy}{dx}\right)^2} \, dx.$$

Because the rightmost integral is being done with respect to x, the lower and upper limits "start" and "finish" are given by $x = 0$ and $x = R$, respectively. As we showed in Chapter 17, if the launch angle and speed are α and v_0 (in Chapter 17 we used θ and V, but that's a trivial notational change), then

$$R = \frac{2v_0^2}{g} \sin(\alpha) \cos(\alpha).$$

So, our problem is find that value of α that maximizes L, where

$$L = \int_0^R \sqrt{1 + \left(\frac{dy}{dx}\right)^2}\, dx.$$

Now, as we found in Chapter 18, the equation for the parabolic path of a football punt is

$$y = x \tan(\alpha) - \frac{g}{2v_0^2 \cos^2(\alpha)} x^2.$$

So,

$$\frac{dy}{dx} = \tan(\alpha) - \frac{g}{2v_0^2 \cos^2(\alpha)} x.$$

I'll let you fill in the details, but if you plug this expression into the L-integral and are careful with the algebra, you will find that

$$L = \frac{g}{v_0^2 \cos^2(\alpha)} \int_0^R \sqrt{\frac{v_0^4 \cos^4(\alpha)}{g^2} + \left\{ x - \frac{v_0^2 \sin(\alpha) \cos(\alpha)}{g} \right\}^2}\, dx.$$

It is immediately clear that

$$\frac{v_0^2 \sin(\alpha) \cos(\alpha)}{g} = \frac{1}{2} R,$$

and with just a bit of easy algebra you should be able to show that

$$\frac{v_0^4 \cos^4(\alpha)}{g^2} = \left\{ \frac{R}{2 \tan(\alpha)} \right\}^2,$$

and so

$$L = \frac{g}{v_0^2 \cos^2(\alpha)} \int_0^R \sqrt{\left\{\frac{R}{2\tan(\alpha)}\right\}^2 + \left\{x - \frac{1}{2}R\right\}^2} \, dx.$$

If we next make the change of variable

$$u = x - \frac{1}{2}R,$$

and so, $dx = du$, then

$$L = \frac{g}{v_0^2 \cos^2(\alpha)} \int_{-R/2}^{R/2} \sqrt{u^2 + \left\{\frac{R}{2\tan(\alpha)}\right\}^2} \, du.$$

This integral is of the general form[3]

$$\int \sqrt{u^2 + a^2} \, du = \frac{u\sqrt{u^2 + a^2}}{2} + \frac{a^2}{2}\ln(u + \sqrt{u^2 + a^2}),$$

where

$$a = \frac{R}{2\tan(\alpha)}.$$

Applying this term to the L-integral, and skipping a few algebraic steps that I'll let you do, we arrive at

$$L = \frac{v_0^2}{g}\left[\sin(\alpha) + \cos^2(\alpha)\ln\left\{\sqrt{\frac{1 + \sin(\alpha)}{1 - \sin(\alpha)}}\right\}\right].$$

Then, "noticing" the identity[4]

$$\sqrt{\frac{1 + \sin(\alpha)}{1 - \sin(\alpha)}} = \frac{1 + \sin(\alpha)}{\cos(\alpha)},$$

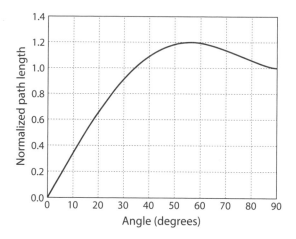

Figure 21.2. The normalized parabolic path length of a punt as a function of α

we finally arrive at

$$
L = \frac{v_0^2}{g}\left[\sin(\alpha) + \cos^2(\alpha)\ln\left\{\frac{1+\sin(\alpha)}{\cos(\alpha)}\right\}\right].
$$

Figure 21.2 shows a plot of $L/\frac{v_0^2}{g}$ versus α (that is, a *normalized L versus* α). As you can see, there is, indeed, a maximum at about $\alpha = 55°$ (a detailed numerical study shows that $\alpha = 56.46°$ is a more precise result[5]). The maximum is quite broad, however, and so the precise value of α is not critical.

It is interesting to compare (for the same v_0) the hang time and the range of the $\alpha = 45°$ punt (the maximum-range punt) to the same quantities for the $\alpha = 56.46°$ punt (the maximum–path length punt). The normalized hang time for all punts is

$$
\frac{T}{v_{0/g}} = 2\,\sin(\alpha),
$$

and the normalized range for all punts is

$$
\frac{R}{v_0^2/g} = 2\,\sin(\alpha)\cos(\alpha),
$$

which gives us the following comparison table.

TABLE 21.1
Hang Time and Range for Two Values of α.

α	*Normalized hang time*	*Normalized range*
45.00°	1.414	1.000
56.46°	1.667	0.921

Thus, the price paid in range by switching to $\alpha = 56.46°$ from $\alpha = 45°$ is a decrease of 7.9%, but the reward is an *increase* by nearly 18% in hang time. The normalized maximum possible hang time (for the worthless $\alpha = 90°$ punt) is 2, and so the $\alpha = 56.46°$ punt achieves more than 83% of the maximum possible hang time while retaining 92% of the maximum possible range.

Who says you can't have your cake and eat it, too?

Notes

1. As I've done throughout our previous analyses of projectile trajectories, I'm ignoring air drag affects. If you are getting tired of reading my disclaimers about air drag, then you *can* read all about how to take it into account (with some significant math complications, because, unlike in this book, the physics *isn't* simple) in my book *Mrs. Perkins's Electric Quilt*, Princeton 2009, pp. 120–135.

2. Generally, I believe punters almost always kick with full strength, with the major exception being an onside kick, in which a *short* kick is the goal for strategic reasons. We aren't studying that situation here.

3. This indefinite integration formula comes from simply looking it up in a table of integrals (see note 4 in Chapter 19). You can, of course, *verify* the formula by differentiation.

4. To show this is a good exercise in algebra, and I encourage you to verify the identity.

5. This numerical result first appeared in Haiduke Sarafian, "On Projectile Motion," *The Physics Teacher*, February 1999, pp. 86–88.

22. Easy Ways to Measure Gravity in Your Garage

Orbits are not difficult to comprehend. It is gravity which
stirs the depths of insomnia."
—*Norman Mailer*, Of a Fire on the Moon *(1970)*

By the time you've reached this chapter you've surely come to expect
to see a g appear in half of all equations. The g, of course, is
the acceleration of gravity at the Earth's surface, equal to about
9.8 meters/seconds-squared $\approx$ 32.2 feet/seconds-squared. When we did
our calculations involving projectile motion, the transit tube, the bungee
jump, and rolling cylinders down an inclined plane, for example, g
invariably popped up somewhere in the math. That happened with
such regularity that it is easy to think g *must* make an appearance when
studying the physics of "things moving through space."

But that's not so. Here's a surprising counterexample, first (as far
as I know) discussed in an elegant 1960 paper.[1] Imagine (as shown
in Figure 22.1) a mass m, initially at rest, sliding down a frictionless
inclined plane through a vertical drop of h. At the bottom of the plane
the mass is launched at ground level into space at speed v_0 at angle α.
What is R, the horizontal distance from the launch point to where the
mass hits the ground?

As we found in Chapter 18, for a projectile launched from the origin
at speed v_0 at angle α, the range is given by (see boxed equation B
there)

$$R = \frac{2v_0^2}{g} \cos(\alpha) \sin(\alpha).$$

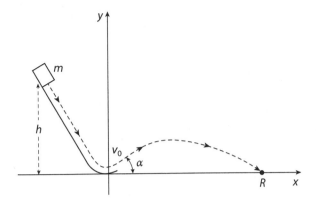

Figure 22.1. R is *independent* of g!

By the conservation of energy, we can write

$$\frac{1}{2}mv_0^2 = mgh,$$

which says the kinetic energy of the mass at launch is equal to the decrease in the potential energy of the mass. That is,

$$v_0^2 = 2gh.$$

Thus,

$$R = 4h\cos(\alpha)\sin(\alpha),$$

an expression, you'll notice, *that has no g in it*. As the author in note 1 wrote, "If this experiment were performed on the Moon, or Mars, with the [mass sliding] down the same incline ... it would [have the same range as it does on Earth]."[2]

What is different on Earth, on the Moon, and on Mars is the launch speed v_0. Where g is greater, the launch speed is greater, which exactly compensates for the increased gravity, to give the same R. The math makes this obvious, but I don't think it is apparent beforehand. Still, despite this pretty little calculation, g does seem to have a habit of appearing in our equations. So, it's important to know its value. And that's our question here, the final one of the book—just how *do* you measure g?

I wrote on this question a few years ago,[3] with the discussion starting as follows:

> The experimental determination of the value of g is, in fact, a classic experiment performed each year in thousands of college freshman physics labs worldwide. I remember well when I did it as a freshman in Physics 51 at Stanford (1958). I remember it as a clunky, uninspiring experiment that required watching a high-speed, pulsed spark generator burn holes through a strip of falling wax paper (I recall that even the graduate student teaching assistant looked like she would rather have been somewhere else). That was followed by the measurement of the distances between adjacent burn holes to eventually arrive, with some arcane intermediate calculations, at a value for g. Here's a better way— faster and pedagogically superior—to measure g. ...All you'll need in the way of equipment is a yardstick, a bouncy rubber ball, and a stopwatch. You don't need an expensive and mysterious (to most college freshmen, anyway) spark generator. You do need to be able to follow a little elementary physics and some simple high school algebra. Then you can measure g where you live in less than 60 seconds.

There then followed a three-page analysis involving some quite simple physics, resulting in this formula for g:

$$g = \frac{8h_0 c^2}{T_n^2}\left(\frac{1-c^n}{1-c}\right)^2,$$

where, if the ball is dropped from a height of h_0, and h_1 is the height of the first bounce, then

$$c = \sqrt{\frac{h_1}{h_0}},$$

and T_n is the time for n bounces (pick any convenient n). The ease of carrying out this procedure is obvious (for *Mrs. Perkins's*, I actually went out to my garage one evening and did it—it *was* easy and a *lot* more fun than the Stanford lab was), and there are other equally easy and, perhaps surprisingly, even *easier* ways of determining g.

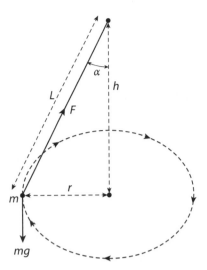

Figure 22.2. The geometry of the conical swing

The rest of this chapter will show you a few of them. Be assured, however, that physicists who are devoted to getting really accurate values of g don't use either the bouncing ball or any of the other methods in this chapter, which are accurate (at best) to only a few percent. However, those physicists spend a fair amount of money on sophisticated equipment,[4] while each of the approaches I'll tell you about here are both easy to do and *inexpensive* (less than \$20).

The Conical Spin

Imagine holding one end of a nearly massless but strong string (a nylon fishing line is a good approximation) in your hand, with the other end tied to a fairly substantial mass (several metal washers wired together should do nicely). Then, as shown in Figure 22.2, move your hand so as to get the weight to swing at a steady speed in a horizontal, circular path of radius r. As shown in the figure, the distance from your hand to the center of the circular orbital plane is h, and the length of the string is L. The tension in the string is F.

We know that the centripetal acceleration experienced by the orbiting mass is $\frac{v^2}{r}$, where v is the speed of the mass. If we write T

as the time for one complete orbit, then

$$v = \frac{2\pi r}{T},$$

and so the centripetal acceleration is

$$\frac{4\pi^2 r^2}{T^2 r} = \frac{4\pi^2 r}{T^2},$$

which means that the inward-directed (radial) force required by this acceleration is

$$m \frac{4\pi^2 r}{T^2}.$$

This force is supplied by the inward-directed (radial) horizontal component of the string tension, which is $F\sin(\alpha)$; that is,

$$F\sin(\alpha) = m \frac{4\pi^2 r}{T^2}.$$

Now, since the orbiting mass has no vertical motion, we know that the net vertical force is zero. That means the upward vertical component of the string tension must exactly balance the downward gravitational force on the mass, and so

$$F\cos(\alpha) = mg;$$

that is,

$$m = \frac{F}{g}\cos(\alpha),$$

and so (not bothering at this point to cancel the F's on both sides), we have

$$F\sin(\alpha) = \frac{F}{g}\cos(\alpha)\frac{4\pi^2 r}{T^2}.$$

(I've put this expression in a box because I'll be referring to it in just a bit.) Finally, from geometry, we have

$$\frac{r}{L} = \sin(\alpha), \quad \frac{h}{L} = \cos(\alpha),$$

and so, substituting these last two expressions into the boxed equation (and at last canceling the F's), we get our result:

$$g = \frac{4\pi^2 h}{T^2}.$$

Notice that we don't need to know m, r, or L, just h and T.

To carry out this procedure by hand, however, clearly requires a pretty steady hand. If you mechanize it just a bit by replacing your hand with the shaft of a vertically mounted synchronous electric motor, it becomes a lot easier to do.[5] Using a 60 rpm motor, for example, automatically gives $T = 1$ second for the period of one orbit, and so now, you don't need a stopwatch. The only measurement left to determine is h. This way of doing the experiment does introduce a curious little twist: it won't work unless L is longer than a certain critical length, although once that critical length is exceeded, it doesn't matter what L actually is! Here's why.

We return to the boxed equation, cancel the F's, and obtain

$$\frac{g}{\cos(\alpha)} = \frac{4\pi^2 r}{T^2 \sin(\alpha)} = \frac{4\pi^2 r}{T^2 \left(\frac{r}{L}\right)} = \frac{4\pi^2 L}{T^2} = \left(\frac{2\pi}{T}\right)^2 L.$$

Writing the *constant* $\frac{2\pi}{T}$—remember, T is now fixed because we are using a synchronous motor—as ω (this is the *fixed angular speed* of rotation of the mass m), we have

$$\frac{g}{\cos(\alpha)} = \omega^2 L,$$

or

$$\cos(\alpha) = \frac{g}{\omega^2 L}.$$

For this to make physical sense (for α to be real) we must have $\cos(\alpha) < 1$, which means that

$$L > \frac{g}{\omega^2}.$$

For a 60 rpm motor ($T = 1$ second) we have

$$L > \frac{32.2 \text{ feet/seconds-squared}}{\left(\frac{2\pi}{1 \text{ second}}\right)^2} = \frac{32.2}{4\pi^2} \text{ feet} = 0.816 \text{ feet},$$

and so L must be longer than just a bit less than 10 inches.[6]

The Horizontal Spin

This second method also involves spinning a mass in a horizontal, circular orbit, using nothing more exotic than a simple tube (the central cardboard tube from a roll of paper towels will do). The setup is shown in Figure 22.3, where you thread the fishing line through the tube and tie equal masses (use washers again) to each end. Then, holding the tube upright in your hand, set the upper mass into a circular orbit of radius r and period T.

The orbital speed is (as with the conical spin)

$$v = \frac{2\pi r}{T},$$

and so the centripetal acceleration is

$$\frac{v^2}{r} = \frac{4\pi^2 r}{T^2},$$

and so the tension in the fishing line is

$$F = m\frac{4\pi^2 r}{T^2}.$$

This tension is provided by the gravitational attraction on the hanging mass, and so

$$F = mg = m\frac{4\pi^2 r}{T^2},$$

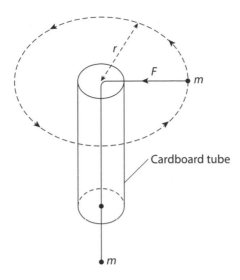

Figure 22.3. The geometry of the horizontal spin

or

$$g = \frac{4\pi^2 r}{T^2}.$$

The originator[7] of this clever method suggested an equally clever way of measuring r: "several knots should be tied in the [fishing line] at known distances from [the orbiting mass] so as to make the radius r easily measurable." That is, simply spin-up the orbiting mass until one (or two or three) premeasured knots just emerge from the tube, and then have a friend use a stopwatch to time the completion of an integer number of orbits to get the average value of T. That's it!

The Vertical Spin

For the next method of determining g using practically nothing, you'll again tie a mass m (again, a bunch of metal washers) to the end of a string and spin it around in a circle of radius r, but now the orbit will lie in a *vertical* plane, as shown in Figure 22.4. You'll spin the mass in a very special way—after you get it going at a pretty good clip, slowly reduce the spin rate until you sense the string *just* going slack when

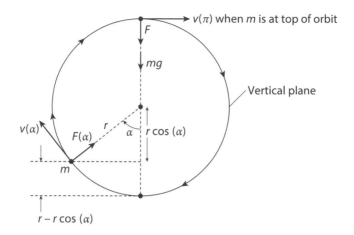

Figure 22.4. The geometry of the vertical spin

the mass is at the top of its orbit. (This can take some trial-and-error practice to get the hang of, but the originator[8] of this method claimed his students soon mastered the technique.) This may strike you as an odd thing to do, but here's why it is key to the method.

Unlike in either the conical or horizontal spin, in the vertical spin the orbital speed of the mass and the tension in the string are *not* constants. Rather, if α is the angle in Figure 22.4 that specifies where the mass is as it orbits, then the speed and the tension are both functions of α. That is, $v = v(\alpha)$, and $F = F(\alpha)$. In particular, $v(0)$ is the speed at the bottom of the orbit, while $v(\pi)$ is the speed at the top of the orbit ($\alpha = \pi \ radians = 180°$). The following analysis is based on the conservation of energy; that is, the sum of the mass potential energy (P.E.) and kinetic energy (K.E.) is, for all α, constant. We'll take the zero reference point for the P.E. as the bottom of the orbit.

At the top of the orbit the K.E. and the P.E. of the mass are

$$\text{K.E.} = \frac{1}{2}mv^2(\pi), \quad \text{P.E.} = 2rmg,$$

while for an *arbitrary* α,

$$\text{K.E.} = \frac{1}{2}mv^2(\alpha), \quad \text{P.E.} = [r - r\cos(\alpha)]\,mg = rmg\,[1 - \cos(\alpha)].$$

So, by conservation of energy, we can write

$$\frac{1}{2}mv^2(\pi) + 2rmg = \frac{1}{2}mv^2(\alpha) + rmg\left[1 - \cos(\alpha)\right],$$

or

$$\frac{1}{2}v^2(\pi) + 2rg = \frac{1}{2}v^2(\alpha) + rg\left[1 - \cos(\alpha)\right].$$

We can determine $v(\pi)$ as follows. The centripetal acceleration of the mass *at the top* of the orbit is given by

$$\frac{v^2(\pi)}{r},$$

which requires the force

$$m\frac{v^2(\pi)}{r}.$$

This force is provided by the sum of the string tension F and the gravitational force on the mass, which are in-line (and, of course, both are directed downward). So,

$$m\frac{v^2(\pi)}{r} = F + mg.$$

Next, since $F = 0$ at the top of the orbit (remember, you are purposely swinging the mass so the string *just* goes slack at the top), we have

$$\frac{v^2(\pi)}{r} = g,$$

or

$$v^2(\pi) = rg.$$

Thus, our conservation of energy equation becomes

$$\frac{1}{2}rg + 2rg = \frac{1}{2}v^2(\alpha) + rg\left[1 - \cos(\alpha)\right].$$

I'll let you do the easy algebra to show that

$$v(\alpha) = \sqrt{3rg\left\{1 + \frac{2}{3}\cos(\alpha)\right\}}.$$

Now, here comes the crucial observation that pushes the analysis forward. If $ds = r\,d\alpha$ is a differential portion of the total orbital path, then the differential time dt it takes the mass to travel that distance is

$$dt = \frac{ds}{v(\alpha)} = \frac{r\,d\alpha}{v(\alpha)}.$$

So, the total time T for one complete orbit (the orbital period) is

$$T = \int dt = \int_0^{2\pi} \frac{r\,d\alpha}{v(\alpha)},$$

where the integration is performed over one orbit. That is,

$$T = \int_0^{2\pi} \frac{r\,d\alpha}{\sqrt{3rg\left\{1 + \frac{2}{3}\cos(\alpha)\right\}}} = \sqrt{\frac{r}{3g}} \int_0^{2\pi} \frac{d\alpha}{\sqrt{1 + \frac{2}{3}\cos(\alpha)}},$$

or, solving for g, we get

$$g = \frac{r}{3T^2}\left\{\int_0^{2\pi} \frac{d\alpha}{\sqrt{1 + \frac{2}{3}\cos(\alpha)}}\right\}^2.$$

The definite integral is, of course, a pure number. The originator (note 8) of this method suggested the integral be evaluated graphically (using the area interpretation of the integral) and stated the result is "roughly 7." In fact, it is easy to show that the integral is an *elliptic integral of the first kind*,[9] with a value that can be looked up in tables (giving 6.993, which *is* pretty nearly 7).

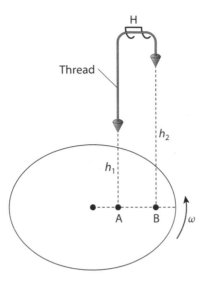

Figure 22.5. The geometry of the double fall

The Double Fall

All the methods for determining g that I've shown you so far involve spinning a mass at the end of a string. This next method, the last of the chapter, gets back directly to what we more commonly associate with gravity: *stuff falling*. You'll also have noticed that with each of the earlier analyses, we went ever further back in time. With this final method we are going all the way back to the late nineteenth century, to an 1884 textbook titled *The New Physics*. Written by John Trowbridge (1843–1923), who was a professor of physics at Harvard from 1870 until his retirement in 1914, it describes a way to measure g that is beautifully elegant in its theory. Figure 22.5 shows the experimental setup.

Hanging above a motionless (for the moment) disk that will, when powered, rotate at a constant angular speed, are two identical, heavy, sharp-pointed plumb bobs. They are positioned so that if dropped, they will hit the disk at the two points A and B, points that are on a common radial line. Imagine, in fact, that a piece of stiff paper is taped to the disk, so that each plumb bob will punch a hole in the paper if dropped. As shown in the figure, the two plumb bobs are connected together by a thread (you'll see why, soon) that passes over a double hook, H, with the plumb bob closest to the disk's center at height h_1 and the other plumb bob at height h_2, with $h_1 < h_2$.

Now, set the disk to rotating at the constant angular speed ω, and then execute a *simultaneous* drop of the two plumb bobs. The word *simultaneous* is the key here, and one easy way to achieve such a drop is to *burn* the thread with a match. This is better than cutting the thread with scissors, as it avoids even the tiny jostling of the plumb bobs that a cutting of the thread by opposing scissor blades almost surely would cause. After the thread parts, our two pointy plumb bobs drop down onto the disk, taking times t_1 and t_2, respectively, where

$$\frac{1}{2}gt_1^2 = h_1,$$

and

$$\frac{1}{2}gt_2^2 = h_2.$$

Clearly, $t_2 > t_1$. That is,

$$t_1 = \sqrt{\frac{2h_1}{g}} < t_2 = \sqrt{\frac{2h_2}{g}}.$$

The difference in these drop times is therefore given by

$$\Delta t = t_2 - t_1 = \sqrt{\frac{2}{g}} \left(\sqrt{h_2} - \sqrt{h_1} \right).$$

When the first plumb bob hits the rotating paper, it punches a hole. When the second plumb bob hits the rotating paper Δt later, it also punches a hole. Since the paper is rotating, these two holes will *not* be on a common radial line. In fact, they will be on two different radial lines that form an angle θ, where

$$\theta = \omega \Delta t = \omega \sqrt{\frac{2}{g}} (\sqrt{h_2} - \sqrt{h_1}),$$

which we can solve for g to get

$$g = \frac{2\omega^2 \left(\sqrt{h_2} - \sqrt{h_1}\right)^2}{\theta^2}.$$

If we know h_1, h_2, and ω, and if we measure θ with a protractor, then we can calculate g.

I don't know what Trowbridge used for the rotating disk, but a modern writer suggests using an old record player turntable.[10] These devices are not as common today as they were in the 1950s (when I was in high school, and they were in the bedroom of every Western society teenager in the world), but they are still around.[11] The standard turntable has three selectable speeds: $33\frac{1}{3}$ rpm, 45 rpm, and 78 rpm. Because even slight deviations from the selected speed would transform a romantic song into one sung by either a chipmunk or by a "voice from deep inside a barrel," turntables are actually impressively accurate in their time keeping.

Once h_1 and h_2 have been selected, the only measurement to be made is that of θ. If, say, we use a 78 rpm turntable, with $h_1 = \frac{1}{2}$ foot and $h_2 = 2$ feet, what value should we expect to see for that angle? Since 78 rpm is

$$\omega = \frac{78}{60} \times 2\pi \, \frac{\text{radians}}{\text{second}} = 2.6\pi \, \frac{\text{radians}}{\text{second}},$$

we have

$$\theta = \omega \Delta t = 2.6\pi \sqrt{\frac{2}{32.2}} \left(\sqrt{2} - \sqrt{\frac{1}{2}}\right) \text{radians} = 1.44 \text{ radians} \approx 82°,$$

an easily measured angle. At turntable speeds of $33\frac{1}{3}$ rpm and 45 rpm, θ would be $\approx 35°$ and $\approx 48°$, respectively.

Now, I'll end this chapter with a historical observation that is not often appreciated, even by many physicists. With the value of g in hand, which this chapter has shown is not difficult to determine, it is now possible to estimate G, the constant in Newton's inverse-square law for the gravitational attraction force F between two point masses M and m, distance r apart:

$$F = G\frac{Mm}{r^2}.$$

It's important to understand that Newton never wrote this equation (and, indeed, never wrote of g![12]); the constants G and g were introduced into physics long after Newton's death. In particular, G as the "gravitational constant" didn't appear until the end of the nineteenth century.

If we take M to be the mass of the Earth and m to be some other mass (say, for example, a teacup), then the gravitational force on the tea cup (what we call its *weight*) is given by mg. That is, since $r = R$ (the radius of the Earth), then

$$mg = G\frac{Mm}{R^2},$$

and so

$$G = \frac{gR^2}{M}.$$

If the average density of the Earth is ρ, then

$$M = \frac{4}{3}\pi R^3\rho,$$

which gives

$$G = \frac{3g}{4\pi R\rho}.$$

(*Note*: The m in $F = mg$ is called the *inertial* mass, while the m in $F = G\frac{Mm}{R^2}$ is called the *gravitational* mass. The equality of these two masses is called the *principle of equivalency*, one of the starting points in the general theory of relativity.)

It had been known by educated people for centuries before Christ that the Earth is a sphere about 4,000 miles in radius.[13] Further, by observing that the Earth's crust has a density about twice that of water, and making the plausible assumption that the Earth's interior is even denser, Newton suggested that the average density is between five and six times that of water.[14] The Earth's average density is what the Cavendish experiment (see note 3 in Chapter 5) measured in 1798, arriving at a value of 5,540 $\frac{\text{kilograms}}{\text{meters cubed}}$, which is right in the middle of Newton's "guesstimate" interval. Substituting all the relevant numbers into the preceding equation for G, then, Newton could have calculated G! Using the midpoint of his estimate for ρ, he could have calculated

$$G = \frac{3 \times 9.8 \frac{\text{meters}}{\text{seconds squared}}}{4\pi \times 4,000 \text{ miles} \times 1,609 \frac{\text{meters}}{\text{mile}} \times 5,500 \frac{\text{kilograms}}{\text{meters cubed}}}$$

$$= 6.6 \times 10^{-11} \frac{\text{meters cubed}}{\text{kilograms} \cdot \text{seconds-squared}}.$$

This is only 1% different from the modern value.

But wait! you object, since I earlier said that Newton never wrote of g, and certainly never stated a value for it, how could he have known of "9.8 meters/seconds − squared"? My point here is that he *could* have known that value *if* he had done one of the experiments in this chapter. Of course, he would have needed a good timing device to do that, an instrument difficult to find in his day. His preserved research notes include comments on the pendulum clocks he used (see Herivel in note 12) in the gravity experiments he did perform.

By not making this calculation of G, Newton missed an opportunity to put yet another gold star by his name (as if he needed another one). Still, while Newton was a genius he was also human and so could make errors just like the rest of us. As a dramatic illustration of this, see the epilogue for "Newton's Gravity Calculation Mistake." Today, a good high school AP calculus and physics student would have no trouble with the problem that Newton stumbled on. In his defense, however, I suspect that the source of his error (still unknown) was merely an arithmetic slip.

Now, finally, here's a little calculation for you to do. The Earth has a mass and diameter that are greater than the Moon's by factors of 81 and 4, respectively. Show that this tells us the acceleration of gravity on the Moon's surface is $\approx \frac{1}{5}g$. (This was dramatically illustrated by the "golf-ball experiment" performed on Moon-to-Earth television by astronaut Alan Shepard during the *Apollo 14* mission of 1971.)

Notes

1. Richard M. Sutton, "Experimental Self-Plotting of Trajectories," *American Journal of Physics*, December 1960, pp. 805–807.

2. A more recent writer aptly called this "a trick of gravity." See Ronald Newburgh, *The Physics Teacher*, September 2010, pp. 401–402.

3. *Mrs. Perkins's Electric Quilt*, Princeton University Press, 2009, pp. 18–23.

4. See, for example, Kurt Wick and Keith Ruddick, "An Accurate Measurement of g Using Falling Balls," *American Journal of Physics*, November 1999, pp. 962–965. The technique described in this paper (accurate to within 0.01%) measures the time lapse between the breaking of two light beams by a falling ball, with air drag taken into consideration, as well. The timing is done electronically, with microsecond precision.

5. For details, see Henry Klostergaard, "Determination of Gravitational Acceleration g Using a Uniform Circular Motion," *American Journal of Physics*, January 1976, pp. 68–69.

6. What happens if L is shorter than this critical length? Simply, the mass will not orbit in a circular path but, rather, will hang straight down and spin around its own body axis. See the paper in note 5 for the not-very-difficult proof of this.

7. Francis Wunderlich, "Determination of 'g' through Circular Motion," *American Journal of Physics*, December 1966, p. 1199.

8. Albert B. Stewart, "Circular Motion," *American Journal of Physics*, June 1961, p. 373.

9. Take a look at box E in Chapter 19, where we first ran into elliptic integrals, in our study of the transit tube. This is now pure math, not physics, but if you're interested, you can convert our integral into the definition form of an elliptic integral of the first kind by doing the following: (1) write $\int_0^{2\pi} \frac{dx}{\sqrt{1+a\cos(x)}}$; (2) make the change of variable $x = 2u$; (3) do the easy algebra and trigonometry to show that

$$\int_0^{2\pi} \frac{dx}{\sqrt{1+a\cos(x)}} = \frac{4}{\sqrt{1+a}} \int_0^{\pi/2} \frac{du}{\sqrt{1 - \frac{2a}{1+a}\sin^2(u)}};$$

(4) set $a = \frac{2}{3}$ and find the value of the integral using a math table.

10. Thomas B. Greenslade, Jr., "Trowbridge's Method of Finding the Acceleration due to Gravity," *The Physics Teacher*, December 1996, pp. 570–571.

11. You can buy new turntables from Amazon for about $80, and I have found used ones on eBay for as low as $15.

12. Newton did, *of course*, understand the concept of the acceleration of gravity and actually performed experiments. His results were of the form of "distance of a fall in one second," however, and not as a value of so many feet/second-squared. He eventually settled on 196 inches in 1 second, which is very close to the correct value (at 32.2 feet/second-squared, a body will fall 193.2 inches during the first second of fall). See John Herivel, *The Background to Newton's* Principia: *A study of Newton's Dynamical Researches in the Years 1664–84*, Oxford University Press, 1965, pp. 186–189.

13. This realization is usually dated to Eratosthenes of Cyrene (276–194 BC). This is the same Eratosthenes who, besides being a director of the famous lost Library of Alexandria, discovered the fundamental technique for identifying prime numbers called the *sieve of Eratosthenes*. The stories of all these events can be found in any good book on the history of mathematics.

14. You can find this suggestion on p. 418 of Andrew Motte's 1729 English translation (from the original Latin, the international scientific publication language of Newton's times) of the *Principia*, published by the University of California Press in 1934.

23. Epilogue
Newton's Gravity Calculation Mistake

A man of genius makes no mistakes. His errors are ... the
portals of discovery.
—*from James Joyce's* Ulysses *(1922)—words that perfectly
describe Newton*

In *The System of the World*, the third book of his 1687 masterpiece,
Principia, Newton gives a dramatic illustration of how weak is the force
of gravity. He asks his readers to imagine two identical spheres, each
1 foot in diameter with a density equal to Earth's average density
(5.5 times that of water). He then claims that if the spheres, each
initially at rest, are "distant but by 1/4 of an inch, they would not,
even in spaces void of resistance, come together by the force of mutual
attraction in less than a month's time. ... Nay, whole mountains will
not be sufficient to produce any sensible effect."[1] Newton provides
no calculations in support of this claim, and, indeed, in whatever
calculations he did do there *had* to be an error. That's because Newton's
claim is not true, and, in fact, it is in error by a *huge* factor. What follows
is a modern calculation of the time for the two spheres to come into
contact.

Figure 23.1 represents Newton's two spheres, centered on the origin,
with their centers initially at $x = -p - \frac{1}{2}s$ and at $x = p + \frac{1}{2}s$, where p is
the radius of each sphere, and s is their initial separation. By symmetry,
if the center of the right-hand sphere is at x, where $0 \leq x \leq p + \frac{1}{2}s$,
then the center of the left-hand sphere is at $-x$. So, writing F as

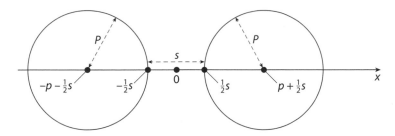

Figure 23.1. Newton's two gravitating spheres, at time $t = 0$

the gravitational attraction force each sphere exerts on the other, by Newton's inverse-square law and his fundamental law of motion (force equals mass times acceleration), we have for the right-hand sphere (for which $x > 0$)

$$F = m \frac{d^2 x}{dt^2} = -G \frac{m^2}{(2x)^2} = -\frac{Gm^2}{4x^2}.$$

The last two terms on the right are negative because the right-hand sphere moves to the left, in the direction of *decreasing* x. Thus,

$$\frac{d^2 x}{dt^2} = -\frac{Gm}{4x^2}, \quad 0 \le x \le p + \frac{1}{2}s.$$

Now, we want to calculate how long it takes x to go from $p + \frac{1}{2}s$ to p, at which point the two spheres just touch.[2]

We'll start the mathematical analysis by switching to the dot notation for derivatives, just as we did in Chapter 19 when we analyzed the high-speed transit tube. That is,

$$\frac{d^2 x}{dt^2} = \ddot{x} = \frac{d\dot{x}}{dt} = \left(\frac{d\dot{x}}{dx}\right)\left(\frac{dx}{dt}\right) = \frac{d\dot{x}}{dx}\dot{x},$$

and so our gravity equation becomes

$$\frac{d\dot{x}}{dx}\dot{x} = -\frac{Gm}{4x^2},$$

which we can rewrite as

$$\dot{x}d\dot{x} = -\frac{Gm}{4x^2}dx.$$

Integrating indefinitely, we get

$$\frac{1}{2}\dot{x}^2 = \frac{Gm}{4x} + C,$$

where C is (for the moment) an arbitrary constant. We can evaluate C if we observe that for the right-hand sphere, $\dot{x} = 0$ when $x = p + \frac{1}{2}s$. This means that

$$0 = \frac{Gm}{4\left(p + \frac{1}{2}s\right)} + C = \frac{Gm}{4p + 2s} + C,$$

and so

$$C = -\frac{Gm}{4p + 2s}.$$

Thus

$$\frac{1}{2}\dot{x}^2 = \frac{Gm}{4x} - \frac{Gm}{4p + 2s},$$

or

$$\left(\frac{dx}{dt}\right)^2 = \left[\frac{Gm}{2x} - \frac{Gm}{2p+s}\right] = Gm\left[\frac{2p+s-2x}{2x(2p+s)}\right]$$

$$= Gm\left[\frac{2\left(p+\frac{1}{2}s-x\right)}{2x(2p+s)}\right] = \frac{Gm}{2p+s}\left[\frac{p+\frac{1}{2}s-x}{x}\right].$$

Solving for $\frac{dx}{dt}$, we obtain

$$\frac{dx}{dt} = -\sqrt{\frac{Gm}{2p+s}\left[\frac{p+\frac{1}{2}s-x}{x}\right]} = -\sqrt{\frac{Gm}{2p+s}}\sqrt{\frac{p+\frac{1}{2}s-x}{x}},$$

where we use the negative square root because we know that the right-hand sphere moves to the left (in the direction of *decreasing x*). That is, for $x < p + \frac{1}{2}s$ the speed of the right-hand sphere is *negative*. Thus, separating variables, we get

$$dt = -\sqrt{\frac{2p+s}{Gm}} \sqrt{\frac{x}{p+\frac{1}{2}s - x}}\, dx.$$

Now, as t varies from 0 to T (the time at which the two spheres just touch) we have x varying from $p + \frac{1}{2}s$ to p. So, integrating, we obtain

$$\int_0^T dt = T = -\sqrt{\frac{2p+s}{Gm}} \int_{p+\frac{1}{2}s}^p \sqrt{\frac{x}{p+\frac{1}{2}s - x}}\, dx.$$

The integral is easy to do. If we set $c = p + \frac{1}{2}s$, then the indefinite integral is

$$\int \sqrt{\frac{x}{c-x}}\, dx,$$

which we can do by first changing the variable to

$$u = (c - x)^{1/2},$$

and so

$$x = c - u^2.$$

Then,

$$\frac{dx}{du} = -2u,$$

and so $dx = -2u\, du$. Thus,

$$\int \sqrt{\frac{x}{c-x}}\, dx = \int \sqrt{\frac{c-u^2}{u^2}}(-2u\, du)$$

$$= -2 \int \sqrt{c - u^2}\, du = -2 \int \sqrt{\left(\sqrt{c}\right)^2 - u^2}\, du.$$

From integral tables, we obtain

$$\int \sqrt{a^2 - u^2}\, du = \frac{u\sqrt{a^2 - u^2}}{2} + \frac{a^2}{2}\sin^{-1}\left(\frac{u}{a}\right).$$

So, since $u^2 = c - x$, and with $a = \sqrt{c}$, we have

$$\int \sqrt{\frac{x}{c-x}}\, dx = -\sqrt{c-x}\sqrt{x} - c \sin^{-1}\left(\frac{\sqrt{c-x}}{\sqrt{c}}\right).$$

Thus,

$$T = -\sqrt{\frac{2p+s}{Gm}}\left[-\sqrt{p+\frac{1}{2}s-x}\sqrt{x} - \left(p+\frac{1}{2}s\right)\right.$$

$$\left.\times \sin^{-1}\left(\sqrt{1-\frac{x}{p+\frac{1}{2}s}}\right)\right]\Big|_{p+\frac{1}{2}s}^{p}$$

$$= \sqrt{\frac{2p+s}{Gm}}\left[\sqrt{\frac{1}{2}s}\sqrt{p} + \left(p+\frac{1}{2}s\right)\sin^{-1}\left(\sqrt{1-\frac{p}{p+\frac{1}{2}s}}\right)\right],$$

or, at last,

$$T = \sqrt{\frac{2p+s}{Gm}}\left[\sqrt{\frac{1}{2}ps} + \left(p+\frac{1}{2}s\right)\sin^{-1}\left(\sqrt{\frac{\frac{1}{2}s}{p+\frac{1}{2}s}}\right)\right].$$

You should verify that that the right-hand-side of this last result is dimensionally correct, that is, has the units of seconds.

For Newton's problem, we have

$$p = \frac{1}{2}\text{ foot} = 0.1524\text{ meters},$$

$$s = \frac{1}{4}\text{ inch} = 0.00635\text{ meters},$$

$$m = \frac{4}{3}\pi r^3 \rho = \frac{4}{3}\pi(0.1524\text{ meters})^3$$

$$\times 5{,}500\frac{\text{kilograms}}{\text{meters cubed}} = 81.547\text{ kilograms},$$

and

$$Gm = 6.67 \times 10^{-11} \frac{\text{meters cubed}}{\text{kilogram} \cdot \text{seconds-squared}} \times 81.547 \,\text{kilograms}$$

$$= 54.4 \times 10^{-10} \frac{\text{meters cubed}}{\text{seconds squared}}.$$

So,

$$T = \sqrt{\frac{0.311}{54.4 \times 10^{-10}}} \left[0.022 + 0.1556 \sin^{-1} \left(\sqrt{\frac{0.003175}{0.1556}} \right) \right] \text{seconds}$$

$$= 335 \,\text{seconds}.$$

Since $\frac{1}{12}$th of a year (Newton's "month") of a 365-day year has 2,628,000 seconds, we see that Newton was in error by a factor of nearly 8,000!

Our final expression for T is, admittedly, fairly complicated. That's because it is an *exact* result, valid for *all* values of s, p, and m, but if we are willing to take advantage of the fact that s is *small* in Newton's problem, then we can get a significantly simpler result. This is a useful check on the exact expression, a technique often used by physicists to gain additional confidence in an exact result. The idea is simple: if two identical point masses are distance r apart, the initial attractive force on each is

$$F = \frac{Gm^2}{r^2}.$$

These attractive forces accelerate the two masses together, and as they approach each other, r decreases and so F increases, and thus the accelerations increase. But if, as in Newton's problem, the distance traveled by each mass compared with r is "small" (just $\frac{1}{8}$ inch for Newton, half of $s = \frac{1}{4}$ inch), then it is a reasonable approximation to take the acceleration as *constant* from start to finish.

From Figure 23.1 the initial separation of the centers of the spheres is

$$r = 2\left(p + \frac{1}{2}s\right) = 2p + s.$$

The initial acceleration is a, where

$$F = ma = \frac{Gm^2}{r^2},$$

and so

$$a = \frac{Gm}{r^2} = \frac{Gm}{(2p+s)^2},$$

which we'll take (as explained above) as constant while each sphere moves through distance $d = \frac{1}{4}$ inch.

The time T' to move distance d, starting from rest, at the constant acceleration a is, as seen earlier in this book, given by

$$d = \frac{1}{2}aT'^2,$$

or

$$T' = \sqrt{\frac{2d}{a}} = \sqrt{\frac{2d}{\frac{Gm}{(2p+s)^2}}},$$

or, since $d = \frac{1}{2}s$,

$$T' = (2p+s)\sqrt{\frac{2d}{Gm}} = (2p+s)\sqrt{\frac{s}{Gm}}.$$

This expression for T' is a *lot* simpler than is the one for T, but recall that T' is an approximation. So, what does T' give, *numerically*?

Using the values for p, s, and m that we calculated earlier, we get

$$T' = (2 \times 0.1524 + 0.00635)\sqrt{\frac{0.00635}{54.4 \times 10^{-10}}} \text{ seconds} = 336 \text{ seconds}$$

which is just *1 second* different from T, our exact answer! T' is, in fact, an *excellent* approximation (for Newton's problem).

This entire discussion is, admittedly, not likely to be a problem that would occur in everyday conversation. Rather, it is the sort of problem that only *physicists* could find enthralling. But I've included it here anyway because one of those physicists happens to be the great Newton, and another (I hope) is you, and also because it is *totally* accessible via simple physics and mathematics.

"Simple physics" doesn't mean simple-*minded* physics, and if this book has convinced you of that, then my job here is done.

Notes

1. You can find this quotation on p. 570 of Andrew Motte's 1729 translation of the *Principia*, published by the University of California Press in 1934.

2. We are using another of Newton's results, too: the gravitational effect of a sphere of uniform density at any point *outside* the sphere is the same as the effect of a *point* mass at the center of the sphere with the same mass as the sphere (see Chapter 5).

Postscript

I guess the thing I missed most in the book, which would
have given a nice overall perspective that would tie together
a number of chapters, is dimensional analysis or
dimensional reasoning. I have found this to work very well
in my teaching at both beginning and advanced levels.
Students enjoy it partly because you can get quite a ways
without doing any detailed calculations.
— *Tom Helliwell, the Burton Bettingen Professor of Physics,
Emeritus, at Harvey Mudd College, in an email to the author after
reading a preliminary draft of this book.*

The epilogue is supposed to be the end of a book, but, well, things
sometimes don't work out as planned. As this book neared completion,
I asked Tom Helliwell, a former colleague of mine when I taught at
Harvey Mudd College (Claremont, California) in the early 1970s, if he
would consider writing a foreword. Of course, you know he said yes
because it's at the front of the book, but Tom didn't just give the book
a quick skim. He didn't just flip through it and then dash off a few nice
words like, "Great book, buy it, you'll like it and, even if you don't like
it, it's big enough to be a good doorstop."

No, Tom *read* it and had a few concerns, none of which I could
squirm out of admitting that he was right. I was able to work all but one
of those concerns into what I had already written, before everything
went to copyediting. But that one exception struck me as *so* central to
what I have tried to do here (I should have thought of it myself) that
I think this postscript will, as Tom suggested, add significantly to the
book.

I am, as you've no doubt guessed from the opening quotation, talking about dimensional analysis. I did address this topic, briefly, in an earlier book, so let me start by repeating what I wrote there:[1]

> When I was a freshman in Physics 51 at Stanford more than 55 years ago, I took a lot of examinations, but one in particular sticks in my memory. One of the questions on that test described a physical situation in which, at the end, the problem was to calculate how far up a glass tube capillary action would draw a fluid. It was a gift question, one the professor had put on the exam to get everybody off to a good start; to answer it all you had to do was remember a formula that had been derived in lecture, and in the course text, and that we had used on at least a couple of homework assignments. All the exam required was plugging numbers into the formula. The professor had kindly provided all the numbers, too. Unfortunately, I couldn't remember the formula and so, no gift points for me.
>
> Later, back in the dorm, I was talking with a friend in the class, who was most grateful for that gift question; he wasn't doing well in the course, and the "free" points were nice.
>
> "So, you remembered the formula, right?" I asked.
>
> "Nope, but you didn't have to. I nailed that one, anyway," he replied.
>
> "What do you mean, you didn't have to remember the formula?" I asked, a sinking feeling in my stomach.
>
> "All you had to do," my friend grinned back, "was just take all the different numbers the prof gave us and try them in different ways until the units worked out as a *length*, the unit of *distance up the tube*." "But, but," I sputtered, "that's, that's . . . *cheating*!"

But, of course, it wasn't cheating. I was just angry at myself for not being sharp enough to have thought of the same idea my friend had. It was my first (painful) introduction to the honorable technique of dimensional analysis. Here, then, as an another example of the idea, is how a physicist might derive the Pythagorean theorem using dimensional analysis.[2]

Figure PS1 shows a right triangle with perpendicular sides of lengths a and b and a hypotenuse of length c. One of the interior acute angles of the triangle is ϕ. I think it obvious that the triangle is determined once we know the values of c and ϕ. That is, for a given c and a given

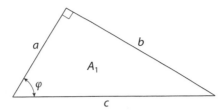

Figure PS1. Deriving the Pythagorean theorem with dimensional analysis (a)

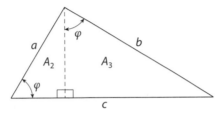

Figure PS2. Deriving the Pythagorean theorem with dimensional analysis (b)

ϕ, the other sides lengths (a and b) and the remaining interior acute angle all have unique values. And certainly, then, the *area* A_1 of the triangle is also determined. Since area has units of length squared, and since ϕ is dimensionless, it must be that the area depends on the square of c. So, let's write the area of our triangle as

$$A_1 = c^2 f(\phi), \tag{1}$$

where $f(\phi)$ is some function of ϕ. (We do not, as you'll soon see, have to know the detailed nature of $f(\phi)$!)

Now, we draw the perpendicular line from the right angle to the hypotenuse of the triangle, as shown in Figure PS2. This divides the triangle into two smaller right triangles, one with area A_2, an acute angle ϕ, and hypotenuse a, and another with area A_3, an acute angle ϕ, and hypotenuse b. Thus, just as in (1), we can write

$$A_2 = a^2 f(\phi) \tag{2}$$

and

$$A_3 = b^2 f(\phi). \tag{3}$$

Since $A_1 = A_2 + A_3$, then $c^2 f(\phi) = a^2 f(\phi) + b^2 f(\phi)$, and so the unknown function $f(\phi)$ cancels (that's why we don't have to know what it is), and we suddenly have, dramatically and seemingly out of thin air, the well-known equation

$$a^2 + b^2 = c^2. \tag{4}$$

That's it. Pretty nifty, don't you think?" (It's easy to show that $f(\phi) = \frac{1}{2} \cos(\phi) \sin(\phi)$—and you should see if you can do this—but the point here is that you don't *have* to show it!)

But that's just *math*, you say—how about another example of dimensional analysis from *physics*, besides my missed opportunity, decades ago, to get some extra test points? Okay, here are *three*. I'll start with the one Professor Helliwell zeroed in on in his email to me, about the opening analysis in Chapter 22. Here is what he wrote: "What could R possibly depend upon, but h, g, m, and alpha? There are no other important parameters in the problem.[3] Alpha is dimensionless, so there is no other parameter that can cancel the dimension of time in g, so g cannot be in the solution. Also, the result cannot depend upon m, because there is nothing to cancel its dimension." So R must be a function of only h and α. To get the specific answer we derived in Chapter 22, however, you really do have to go through the detailed analysis. But Tom is right—the absence of g in this case is, from dimensional analysis alone, no surprise.

For a second physics example, consider sand falling through a circular orifice, as in an hourglass. Made of many tiny, individual solid particles, sand flows through the orifice as would a liquid. But not *like* a liquid. Because of the friction between the individual particles themselves, and the outer particles and the walls of the hourglass, the flow rate (mass per unit time) is observed to be nearly constant—a nice characteristic for a timing device! This means, in particular, that the flow rate doesn't depend on the height of the column of sand waiting to pass through the orifice. This is quite different from the behavior of water flowing out a hole in the bottom of a bucket, for example, where the flow rate *does* depend on the "water head." The only parameters left that could play a role in the flow rate are, then, the sand density ρ, the orifice diameter D, and the acceleration of gravity g.

So, let's write

$$\frac{dm}{dt} = f(\rho, D, g), \tag{5}$$

where f is some function. The word *some* leaves a lot of wiggle room; what *can* we say about f? What we'll do is take f as a *product of powers of the variables*, that is, as

$$f(\rho, D, g) = K\rho^a D^b g^c,$$

where K, a, b, and c are dimensionless constants. We'll do this because, whatever f is, its *functional* dependence on the variables should be independent of the particular choice we make of the units in which to measure length, time, and mass. Nature, after all, doesn't care if we measure in inches or meters, in seconds or days, or in grams or pounds.

To see that our assumed functional form for f has this property, notice that if we measure mass in one system of units, and then again in a new system with a unit that is x times larger, the new mass measurement will be $1/x$ times the first. In the same way, if we let y and z be how many times larger the length and time units are, respectively, in the new system, those new measurements will also be similarly related to the old measurements by the factors $1/y$ and $1/z$, respectively. Writing f as a product of powers preserves the functional dependency and merely shifts the effect of changing units into the constant K. So, since we know there will be a K involved, let's just ignore it for now and concentrate on the functional form of f. When we are done determining that, we'll just insert a K with the understanding that its value (determined by experiment) depends on the system of units we happen to use.

If we write M, L, and T for the dimensions of mass, length, and time, respectively, then on the left-hand-side of (5) the units are $\frac{M}{T}$.

Since ρ, D, and g have the units of $\frac{M}{L^3}$, L, and $\frac{L}{T^2}$, respectively, then with a, b, and c constants, we must have

$$\frac{M}{T} = \left(\frac{M}{L^3}\right)^a (L)^b \left(\frac{L}{T^2}\right)^c = \frac{M^a L^{b+c-3a}}{T^{2c}}.$$

That is, $a = 1$, $b + c - 3a = 0$, and $2c = 1$. These relations easily reduce to $a = 1$, $b = \frac{5}{2}$, and $c = \frac{1}{2}$. Thus, with K some constant (which can be found by experiment),

$$\frac{dm}{dt} = K\rho g^{1/2} D^{5/2}. \tag{6}$$

The exponent of D is probably a big surprise. If the flow rate depended on the area of the orifice, an entirely reasonable *first guess* assumption, the exponent would be 2, not 2.5. But actual experiments with sand flowing through various sizes of orifices show that (6) *is*, in fact, correct.[4]

My final, quite dramatic example of dimensional analysis in physics comes from a real-life story from World War II. In 1941 the English mathematical physicist Sir Geoffrey Taylor (1886–1975) was told of the *possibility* of a superbomb, and he was asked by the British military authorities to think about the physics of a *really big* explosion. This he did in very spectacular fashion, but it wasn't until nearly 10 years later that the world outside the top secret circles of weapons research realized just how far simple physics could take one.

When the first atomic bomb, a plutonium implosion device,[5] was detonated July 16, 1945, in the Alamogordo, New Mexico, desert, that historic event (code named *Trinity*) was filmed by a high-speed (10,000 frames/second) motion picture camera. In 1947 that film was declassified, and individual frames showing the expanding fireball were published worldwide. Each frame was conveniently marked with the instantaneous radius of the nearly perfect hemispherical[6] fireball, along with the time measured from the instant of detonation. One bit of information was *not* declassified, however, and that was the energy of the explosion. *That*, U.S. authorities decided, would remain top secret. So, it was a great surprise for those authorities when, in 1950, Taylor used his 1941 theoretical dimensional analysis, combined with the published photographs from 3 years before, to accurately calculate the energy of the bomb. Here's how he did it.[7]

Arguing that the radius R of the fireball would be a function of the explosion energy E_0, the density ρ of the air the fireball will expand into, and the time t since the instant of detonation, Taylor wrote

$$R = f(E_0, \rho, t) = K E_0^a \rho^b t^c. \tag{7}$$

Setting the dimensions on the left of (7) equal to the dimensions on the right (remember, K is dimensionless, although its value does depend on the system of units used), we have[8]

$$L = \left(\frac{ML^2}{T^2}\right)^a \left(\frac{M}{L^3}\right)^b T^c = M^{a+b} L^{2a-3b} T^{c-2a}.$$

That is, $a + b = 0$, $2a - 3b = 1$, and $c - 2a = 0$, a system of equations easily solved to give $a = 1/5$, $b = -1/5$, and $c = 2/5$. So,

$$R = K E_0^{1/5} \rho^{-1/5} t^{2/5},$$

or, since Taylor had experimental evidence that for the MKS system of units (meters, kilograms, seconds) the value of $K \approx 1$, we have

$$R = \left(\frac{E_0}{\rho}\right)^{\frac{1}{5}} t^{\frac{2}{5}}. \tag{8}$$

Given the relative ease with which we've developed (8), there really is an amazing amount of information in it. For example, if we build two bombs with one having five times the explosive energy of the other, then at any given time after detonation (for a fixed air density) the bigger bomb's fireball will *not* be five times larger but, rather, will be larger by a factor of "only" $5^{1/5} \approx 1.38$. Or if a bomb explodes at a high altitude where the air density is just one-third of what it is at the ground, the resulting fireball at any given time after detonation will *not* be three times that of the ground detonation fireball but, rather, will be larger by a factor of just $3^{1/5} \approx 1.24$.

To see if (8) really did describe what happened in the 1945 explosion, Taylor took the logarithm of both sides and wrote

$$\log_{10}(R) = \frac{1}{5}\log_{10}\left(\frac{E_0}{\rho}\right) + \frac{2}{5}\log_{10}(t), \qquad (9)$$

which says that a plot of $\log_{10}(R)$ versus $\log_{10}(t)$ should be a straight line with a slope of 2/5. When Taylor plotted the radius and time legends given on the declassified bomb film images he got a *virtually perfect* straight line. As he wrote in his paper, "The ball of fire did therefore expand very closely in accordance with the theoretical prediction made more than four years before the explosion took place." This excellent agreement between theory and experiment was quite remarkable, as the range of values for both R and t were "big," namely, $11.1 \leq R \leq 185.0$ (in meters) as t varied (in seconds) over the interval $0.0001 \leq t \leq 0.062$.

Now, how did Taylor get the value of E_0? Writing (9) as

$$5\log_{10}(R) - 2\log_{10}(t) = \log_{10}\left(\frac{E_0}{\rho}\right),$$

we can plug in any pair of values for R and t that were given in the detonation photos. So, using (for example) $R = 185$ meters at $t = 0.062$ seconds, we arrive at

$$\log_{10}\left(\frac{E_0}{\rho}\right) = 13.75,$$

and so

$$E_0 = \rho 10^{13.75}.$$

Taylor used $\rho = 1.25$ kilograms/cubic meter for the density of air, and so

$$E_0 = 1.25 \times 10^{13} \times 10^{0.75} \text{ joules} = 7.03 \times 10^{13} \text{ joules},$$

where we know the units of E_0 are the MKS units of energy (joules) because all the other quantities are in MKS units.

The convention is to report the energy of an atomic explosion (what weapons engineers call the "yield") in units of *metric tons* of

TNT, and that's what Taylor did. (A metric ton is 1,000 kilograms $\approx 2,200$ pounds.) Since 1 pound of TNT releases 1.9×10^6 joules of energy, then 1 metric ton releases 4.18×10^9 joules, and so

$$E_0 = \frac{7.03 \times 10^{13}}{4.18 \times 10^9} \text{ metric tons of TNT} = 16{,}818 \text{ metric tons of TNT.}$$

This is very nearly the number that appears in Taylor's paper (16,800 tons). It is so close to the value the U.S. authorities thought to be the true, *classified* value of the Trinity bomb energy that, for a while, it was thought Taylor had breached military security.

But he hadn't. It was all "just simple physics."

Notes

1. *Mrs. Perkins's Electric Quilt*, Princeton University Press, 2009, pp. 13–15. Dimensional analysis has been around in physics for a long time; it is usually dated from an 1863 paper by the great Scottish physicist James Clerk Maxwell (1831–1879), but you can find a hint of it in Newton's writings.

2. I came across this derivation while browsing in a book by A. B. Migdale, *Qualitative Methods in Quantum Theory*, W. A. Benjamin, 1977 (originally published in Russian in 1975).

3. Well, you might reply, how about v_0, the launch speed? That does *not* count as a parameter, though, because in this problem it, too, is totally determined by h, g, and m. It *was* a parameter in earlier problems, like shooting a bullet out of a gun, since then v_0 didn't depend on *just m* and *g* but also on the amount of gunpowder used.

4. Metin Yersel, "The Flow of Sand," *The Physics Teacher*, May 2000, pp. 290–291.

5. The first atomic bomb ("Little Boy") used in war, dropped on Hiroshima, Japan, was a uranium *gun bomb* (two individually subcritical U-235 masses were slammed together, with one fired from a cannon into the other, to rapidly form a greater-than-critical mass). Scientists were so sure this would work they didn't even bother to test the design. The second bomb dropped on (Nagasaki) Japan ("Fat Man") was an immensely more complicated *implosion bomb* (a subcritical spherical mass of plutonium was suddenly compressed into criticality by the inward moving shock waves generated by a coating

of simultaneously detonated high-explosive "lenses" on the surface of the sphere).

6. *Hemispherical* because the bomb was detonated just 100 feet above ground level (at the top of a tower). A high-altitude bomb detonation would, of course, produce a spherical fireball.

7. Sir Geoffrey Taylor, "The Formation of a Blast Wave by a Very Intense Explosion (part 2): The Atomic Explosion of 1945," *Proceedings of the Royal Society of London A*, March 22, 1950, pp. 175–186. Part 1 of Taylor's two-part paper contains his theoretical 1941 work, on pp. 159–174. Part 2 reproduces some of the declassified fireball photos used by Taylor.

8. To see that the units of energy are $\frac{ML^2}{T^2}$, recall that energy = force × distance = mass × acceleration × distance, and so energy has the units $(M)\left(\frac{L}{T^2}\right)(L) = \frac{ML^2}{T^2}$.

Acknowledgments

There were many people who helped me get this book into your hands.

I did much of the early literature searching in the Physics Library at the University of New Hampshire, and physics librarian Heather Gagnon was of great assistance during the many days I sat in the stacks reading. After leaving the library and walking over to the student union, the staff at the campus Dunkin' Donuts kept me full of coffee (and so awake) as I wrote.

The people at Princeton University Press were, of course, *absolutely crucial*, starting with my super, long-time editor Vickie Kearn and her assistant Betsy Blumenthal, the book's ever-efficient production editor Deborah Tegarden, and the Press's talented artists—Dimitri Karetnikov and Carmina Alvarez-Gaffin—who transformed my amateurish attempts at drawing a straight line into, literally, works of art.

The book's copyeditor, Barbara Liguori in Tucson, Arizona, kept me from looking like someone who slept through high school English (which, alas, I think I actually did now and then). Barbara also reminded me why driving in Arizona and California can be a risky business (see the finally commentary in the lone endnote to Chapter 2).

Three anonymous reviewers provided many helpful suggestions.

My former colleague at Harvey Mudd College, Claremont, California, physicist Tom Helliwell graciously agreed to write a foreword and vigorously refused to accept any payment except my thanks and a copy of the book.

Finally, my wife of 54 years, Patricia Ann, has always supported my writing life, and while I'll never stop trying, I know I will never be able to thank her enough.

Paul Nahin
Lee, New Hampshire
August 2015

P.S. I gratefully thank Anne Karetnikov for the wonderfully exotic cat in the book's cover art. The look on that cat's face mirrors, *exactly*, the spirit in which this book was written.

Index